MY BEST PUZZLES IN MATHEMATICS

By

HUBERT PHILLIPS

(*"Caliban"*)

DOVER PUBLICATIONS, INC.

NEW YORK

Published in the United Kingdom by Constable and Company Limited, 10 Orange Street, London, W.C.2.

My Best Puzzles in Mathematics is a new work published in collected form for the first time in 1961 by Dover Publications, Inc.

Standard Book Number: 486-20091-4
Library of Congress Catalog Card Number: 61-66241

Manufactured in the United States of America

Dover Publications, Inc.
180 Varick Street
New York 14, N.Y.

PREFACE

There is little that I need say by way of introduction to this collection of puzzles. They are all strictly "mathematical," though the mathematical knowledge necessary for solving them is, in all but a handful of cases, elementary. That is because these puzzles have all been published serially in various newspapers and magazines, and their appeal would be negligible if their solution demanded advanced mathematics.

The journals in which they were originally published are the *Daily Telegraph*, the *Evening Standard*, *Truth*, and the *Law Journal*.

They appear here, with their original wording unchanged, except that, where puzzles involve monetary calculations, I have converted pounds, shillings, and pennies into dollars and cents. The factual basis of many of these puzzles is derived from various games, e.g., contract bridge. A good many are based on Britain's national game: association football ("soccer"). You don't need to know this game to tackle the puzzles based on it. Its scoring is simplicity itself: each side, if it scores at all, merely scores one, two, or more goals.

London
August, 1960

HUBERT PHILLIPS

CONTENTS

vii

Contents

Contents

Contents

PUZZLES

1 CINDERELLA

Cinderella met her Good Fairy in the wood, and the Good Fairy (to cut a long story short) said:

"Well, my dear, and what can I do for you?"

"I should like," answered Cinderella, "the gift of perpetual youth."

"And you shall have it," said the Good Fairy.

"You see," Cinderella explained, "I should hate to be as old as Begonia." (Begonia was the elder of her two sisters.) "Begonia's eight years older than I am."

The Good Fairy touched Cinderella with her wand. "Be forever," she said, "your present age."

When her sisters heard about the Good Fairy's gift, they were very, very angry. "You selfish little slut," said Begonia. "You forget that every year Father gives us—to be divided equally among us—a check for that number of thalers which is the product of our three ages in years."

"That's right," said Fuchsia, who had cauliflower ears but was good at mental arithmetic. "Let me see—in the next two years alone your selfishness will cost us 1,382 thalers."

How old is Cinderella?

2 MIRANDA TURNS THE TABLES

"I suppose, Miranda," I said, "that you're quite a whale on algebra now you've passed the School Certificate. I hear you got a Credit in Mathematics."

"I surely did," said Miranda. "In fact, I can turn the tables on you, Caliban. I made up a problem about our ages—mine and Stella's and Eva's and Lucinda's and Dorothea's—and Father was completely stumped by it, though he's every bit as clever as you are."

She showed me her problem, which ran as follows: "The sum of our ages is five times my age. (By 'age' throughout this problem is meant 'age in years.') When Stella is three times my present age, the sum of my age and Dorothea's will be equal to the sum of the present ages of the five of us; Eva's age will be three times her present age; and Lucinda's age will be twice Stella's present age, plus one year."

How old is Miranda?

3 FUN ON THE *STYGIAN*

Allblah, Biffins, Chump, Dimwits, and Effish—all enthusiastic solvers of conundrums—are fellow passengers on the SS. *Stygian*. They therefore agree to amuse themselves as follows:

(1) Each day one of them (the five taking it in turn) sets five conundrums for the other four to solve. The first to solve each conundrum scores one point. In no case can a point be divided; should two solvers claim to finish simultaneously, the point is awarded in accordance with the setter's decision. Five points, and no more, are awarded each day.

(2) At the end of each day each of the four solvers competing that day becomes entitled to one dollar per point in respect of the difference in points between his score for the day and the score for the day of each lower-ranking competitor. The money owing from each solver who, on balance, has lost is paid each day into a pool from which those who, on balance, have won then draw their winnings.

The order of capacity, as solvers, of the five participants is: (1) Allblah, (2) Biffins, (3) Chump, (4) Dimwits, (5) Effish. None of these is ever beaten, in respect of the number of conundrums solved in a single day, by a competitor ranking lower in the order of capacity.

The sums paid into the pool on the five days total $41 in all.

At the end of these five days Allblah has won $20; Dimwits has lost $12; Effish has lost $12.

How much, on balance, has been lost or won by Biffins and Chump respectively?

4 VILLAGE BAZAAR

"Seventy dollars," said the Rector. "That's what I want you children to spend at the Bazaar.

"I'm told that there are five stalls, at which the prices are 1, 2, 3, 4, and 5 dollars respectively. I want each of you to bring back three presents for your mother."

"From different stalls?" asked Margaret.

"Not necessarily," said the Rector. "But each of you must patronize two stalls; and you must not, collectively, patronize more than three stalls in all. Also you must so divide the money that no two of you spend the same amount and that no one spends as much as fourteen dollars."

How many children went to the Bazaar? and which stalls did they patronize?

5 THE JONES BOYS

Last Christmas I was staying with the Jones family. They have three boys: Ifor, Evan, and David. The eldest boy, Ifor, had recently won a scholarship to a well-known public school.

I remembered their telling me, on a previous occasion when I spent Christmas with them, of the principle on which they were given money to spend on Christmas presents. "Daddy takes as a basis" (David, the youngest and brightest, had explained) "the age of each of us in years. Then each of us is given, to spend on the other two, that number of dollars which is the product of their two ages in years."

"And do you still get money on that basis?" I asked them, this time.

"Oh, yes," said David. "This Christmas we are getting, among us, $120 more than we had on the occasion of your last visit."

The boys were born in three consecutive Novembers.
How many years elapsed between my two visits?

6 APRIL'S BROOD

April writes to May as follows:

"So many thanks, darling, for the $200 which you sent me. I divided that amount equally among the boys, and told them to buy presents for their sisters with what I gave them. Each of them bought something for each girl. They all spent their money at the dollar bazaar—you know the place I mean—at one stall everything costs one dollar; at another, two dollars; at another, three dollars; and so on.

"Rather funny, none of the boys bought more than one present at one stall, and no two of them went to the same assortment of stalls. This you wouldn't think possible, but Harold tells me it could just be managed."

How many children has April?

7 INTELLIGENCE TESTS

"Rather fun, Miss Queerie's intelligence tests," said Joy. " 'Seventeeners' she called them. We had a lot of them the fortnight we were in camp."

"Why 'seventeeners'?" I asked.

"Why," said Joy, "for some obscure reason the winner of each test received seventeen cents. But each of us had to lose one cent for every test she didn't win."

"At that rate," said I, "some of you might have had to fork out."

"We did," said Joy. "I, for instance, lost thirty cents on balance. Still, I wasn't dissatisfied. Miss Queerie, who made up the difference between our winnings and our losses, contributed $3.60. It was an interesting affair; each of us won at least one test, but no two of us won the same number."

How many tests did Joy win?

8 BRIDGE

"Bridge last night," said the Doctor. "We played five rubbers, cutting in each rubber. The scores ran rather high: 1,800, 900, 2,300, 1,300, 1,600. We were playing one dollar per hundred."

"Any luck?"

"The Colonel and I won. The Padre and the Admiral lost. The Colonel was the big noise, though. His winnings exceeded mine by fifteen times the difference between the Padre's losses and the Admiral's. The Admiral, by the way, was the bigger loser."

Who partnered whom, and who were the winners, in each of the five rubbers?

9 DODECAHEDRA

I have an indefinite number of regular dodecahedra, indistinguishable in appearance from one another. I have pots of red and blue paint. If each face of each dodecahedron is to be painted red or blue, **how many dodecahedra which are distinguishable from one another shall I be able to produce?**

10 THE SIMIAN LEAGUE

"This year," said Orang, the Hon. Secretary of the Simian League, "the number of clubs in our League has been increased."

"More work for you," I said.

"Definitely. Each club in our League plays home and away matches with each of the others. So that this year the total of matches played is increased by a number equal to sixteen times the number of clubs added to the League."

I went into the question more closely, and made another interesting discovery: that the number of matches played last year, plus the number to be played this year, totaled four less than would have been played this year had yet one more club been added to the League.

How many clubs are there in the Simian League?

11 THE CROSS-COUNTRY FINALS

"I've often thought," said Sprintwell, "that the Cross-Country Finals at Blenkinsop would provide material for a problem."

"Do tell me about them."

"Well," said Sprintwell, "this is how they were organized. Ten chaps were chosen for the finals—one from each House— and each of them had to do the best he could, because he was running for his House. None of your first, second, and third business. Every runner scored points because his final score might be negative—according to the Smith-Dingo formula; and what he won—or lost—was clocked up to the account of his House in the final reckoning for the Inter-Athletic Trophy."

"I follow you so far," said I.

"I should jolly well hope you did," said Sprintwell. "Well, now: about these points. The ten selected runners ran over a selected course, and at the end of the run they were divided into three classes. Class A runners were those who had covered the course in less than so many minutes: Class B runners had taken less than so many more minutes; the remaining runners went in Class C. No one knew what the times in question were until after the race; so we all had to sprint as hard as we jolly well could. And now for the marking—the Smith-Dingo formula—I'm sure you'll agree it's ingenious.

"This was it. A chap in Class A scored 2 points in respect of each runner in Class B and 5 in respect of each runner in Class C. A chap in Class B scored 2 points in respect of each runner in Class C and lost 3 points in respect of each runner in Class A. And a chap in Class C—well, he got it in the neck. He lost 2 points in respect of each runner in Class B and 5 points in respect of each runner in Class A."

I pondered. "On that basis," I said, "the net aggregate points scored by the ten runners must generally have been negative."

"Why, yes," said Sprintwell, "they were. Jolly bright of you to see it so quickly, though. I ran three times for my House, Dickery's—and, on balance, taking the three years together, 17 points were lost. I've forgotten, though, how many were lost in each of the three several years. I dare say you can work it out."

"I should think I could," said I. "Sounds like very inadequate data. Can you remember, by any chance, how many points you scored each year? Or perhaps that's not a tactful question?"

"Oh yes, it is," said Sprintwell. "I made a plus score each year—and the same plus score each year, too."

What was Sprintwell's plus score each year?

12 GOOD EGGS

"You don't like arithmetic, child?" said Humpty Dumpty. "I don't very much."

"But I thought you were good at sums," said Alice.

"So I am," said Humpty Dumpty. "Good at sums; oh, certainly. But what has that to do with liking them? When I qualified as a Good Egg—many, many years ago, that was—I got a better mark in arithmetic than any of the others who qualified. Not that that's saying a lot. None of us did as well in arithmetic as in any other subject."

"How many subjects were there?" said Alice, interested.

"Ah!" said Humpty Dumpty, "I must think. The number of subjects was one-third of the number of marks obtainable in any one subject. And I ought to mention that in no two subjects did I get the same mark, and that is also true of the other Good Eggs who qualified."

"But you haven't told me——" began Alice.

"I know I haven't," said Humpty Dumpty. "I haven't told you how many marks in all one had to obtain to qualify. Well, I'll tell you now. It was a number equal to four times the maximum obtainable in one subject. And we all just managed to qualify."

"But how many——" said Alice.

"I'm coming to that," said Humpty Dumpty. "How many of us were there? Well, when I tell you that no two of us obtained the same assortment of marks—a thing which was only just possible—you'll be well on the way to the answer. But to make it as easy as I can for you, I'll put it another way. The number of other Good Eggs who qualified when I did, multiplied by the number of subjects (I've told you about that already), gives a product equal to half the total number of marks obtained by each Good Egg. And now you can find out all you want to know." He composed himself for a nap.

Alice was almost in tears. "I can't," she said, "do any of it. Isn't it differential equations, or something I've never learned?"

Humpty Dumpty opened one eye. "Don't be a fool, child," he said crossly. "Anyone ought to be able to do it, who is able to count on five fingers."

What was Humpty Dumpty's mark in arithmetic?

13 GRINDGEAR'S REGIS-TRATION NUMBER

When Professor Grindgear went to the Police Station to report the loss of his new car, he could not remember its number.

"Four figures, Inspector," he said. "I know that, but I can't remember a single one of them."

"Four figures, eh?" said Sergeant Smallbeer. "Well, sir, that's something. But we'd like a more definite clue, if you could give us one."

Grindgear's blue eyes assumed a glazed expression. "Wait," he said. "Something begins to come back to me. The other day, apropos my car number, I was discussing factors with my granddaughter."

"Factors, eh?" said Smallbeer, moistening his pencil.

"Yes," said the Professor, "my car number has quite a lot of factors. And I was mentioning to Cyclamen (my granddaughter —a silly name, isn't it?)—I was mentioning to her that three of these factors, adding up to 100, could be selected in quite a lot of ways. See what I mean? Suppose my car number is n——"

"n," said the bewildered Smallbeer, dutifully making a note.

"—suppose my car number is n. If ax, by, and cz are each equal to n, then a, b, and c can be selected in such a way that $a + b + c = 100$, $x + y + z$ is a minimum. . . . Yes, I remember it now. . . . In that case, we discovered, $x + y + z$ is equal to the product of four numbers, all primes; and the sum of those four numbers—this is what we thought rather interesting—is one more than a and one less than b. And that fact, I'm pretty sure, affords a unique clue to the number of my car."

"And what did you say it was, sir?" asked Smallbeer, who had been making out a football coupon during the latter part of the Professor's discourse.

"Pooh!" was the disconcerting answer. "Work it out for yourself, Smallbeer. I've given you all the data."

What is the Professor's car number?

14 THE SOCCER CHAMPIONSHIP

Anderson, Barnes, Chaplin, Dickens, Egerton, and Fowler are all very keen on soccer. They decided to hold an individual championship. They therefore held a series of matches, in the course of which they divided themselves into two teams of three players each in every possible way.

A player scored 2 points for a match in which his team won, one for a match in which it drew; the championship was to go to the player with most points. Where points were equal, the player whose teams had scored more goals was to be placed higher.

Dickens won the championship after an exciting series of games. Yet his teams scored fewer goals than their opponents. Only 23 goals were scored in all; Fowler was bottom of the table, though he had a balance of goals in his favor.

Only two matches had the same result.

Second place was taken by Anderson, whose teams scored the greatest number of goals. The other three competitors were bracketed in the final table. Barnes's teams, however, scored goals in more matches than did Egerton's, but in fewer matches than Chaplin's.

Give full particulars of each match played, other than those which were drawn.

15 TRIANGLE GOLF

Slicer, Divot, and Fluff spend a golfing holiday at North Hoyland. Each morning, as a prelude to more serious business, they play a three-ball match over the first five holes. It is agreed, for the purposes of this match, that there shall be a defined winner at each hole (if two players take the same number of strokes the hole is played again). Stakes are paid after five holes, on the following basis:

A player who wins a hole receives $1 from each of the others.

A player who wins two holes in succession receives $3 ($2 + $1) from each of the others.

A player who wins three holes in succession receives $6 ($3 + $2 + $1) from each of the others.

Similarly, a player receives $10 for four holes won in succession and $15 for all five holes.

Stakes are settled according to differences in the score at the end of the five holes.

This little match is played for eight successive days, at the end of which the position is as follows:

(1) Slicer has lost each day; and each day more than the day before. He won the fourth hole the first day, the first hole the second day, and the fifth hole the third day.

(2) Divot won the third hole the third day.

(3) Fluff won two holes (and two holes only) each day. One of the holes he won was the same on seven days out of the eight.

Name the two holes which were won by Fluff on each of the eight days.

16 TALL SCORING

The four Houses at Midsummer Preparatory School each played one match at soccer against each of the others last season. There was some tall scoring, as witness the final table:

	GOALS FOR	GOALS AGAINST	POINTS SCORED
Snug	13	17	4
Bottom	17	13	3
Flute	17	13	3
Quince	13	17	2

Two points were scored for a win and one for a draw. Each match produced the same number of goals; no two matches produced the same score. Of their 13 goals, Quince scored two against Flute.

What was the result of the match between Quince and Bottom?

17 BOWLS

"You can settle this one for us, Doctor," said a colleague of Dr. Dingo's at the Club. "Drake here, and his pal Raleigh, have arranged to play a series of matches at bowls." He named the number.

"They will toss each time to decide who first throws the jack. And the question that has cropped up is: what are the odds against Drake's winning the toss on at least four occasions?"

"That's certainly not much of a poser," said Dingo. "It's an even-money chance."

How many matches are Drake and Raleigh to play?

18 MRS. COLDCREAM OBJECTED

"I object," said Councillor Mrs. Coldcream. "I see no reason why the boys should be so favored at the expense of the girls."

This was at a meeting of the Holmshire Education Committee. It had been proposed to award 19 scholarships to boys and girls of the county. Their total value was $1,000; each girl was to receive so many dollars exactly, and each boy $30 more than each girl.

Mrs. Coldcream pressed her point with such fervor that it was decided to reallocate the money. Each girl would receive $8 more than the committee had originally recommended. The boys' scholarships were scaled down accordingly.

How much did each boy, and each girl, receive?

19 FOOZLEDOWN

At Foozledown—famous for its gorse-studded links—there is, as it were, a club within a club. This consists of a small group of golf maniacs who are forever pitting themselves one against another.

A question arose about a year ago as to which two players,

of this group, would best acquit themselves in a foursome. It was eventually decided that each pair should play a match against every other pair.

As a result of this exhausting contest—which made it necessary to play no fewer than 45 matches—Deadpan and Backspin emerged as the triumphant pair.

How many other players took part in the contest?

20 CUBES

I have a number of wooden cubes, of the same size, which are painted white.

I propose to have painted, on each face of each cube, a black line connecting the center points of two opposite edges (as shown in the diagram).

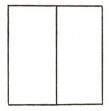

It will, I think, tax all your ingenuity to discover **how many cubes can be produced in accordance with this formula which are distinguishable from one another.**

21 MORE CUBES

In the last puzzle, each face of the cube had painted on it a line joining the central points of two opposite edges; and I showed how many distinguishable cubes could be produced.

Now consider the case where each face of the cube had a line running across it diagonally.

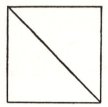

Can you now determine how many cubes, distinguishable from one another, can be produced in accordance with this formula?

22 TABLE TENNIS

Springer, Lush, Wallaby and Cachou competed for a table-tennis trophy. Each of these four had a 50-50 chance of winning a game against any of the others. It was arranged that each should play each of the others twice, the trophy going to the player who won most games. If two or more players won the same number of games, they were to hold the trophy jointly.

Faredooze, the club bookmaker, offered to lay bets on this contest. At the end of the first round (when each player had played each of the others once), Springer had won all his games, Lush had won two, Wallaby had beaten Cachou. Wallaby rang up Faredooze and told him the state of the score. "What will you lay me," he asked, "against my winning the trophy outright?" Faredooze considered. "Fifties." "Fifty to one in dollars?" "That's right," said Faredooze.

Is this a good bet from Wallaby's point of view?

23 KINDHARTZ

Kindhartz Ltd. (manufacturers of coronets) run a non-contributory superannuation scheme for their small but highly skilled staff. Each employee's pension depends on the age of his retirement. He becomes eligible at the age of 50, but can go on working until he is 80.

The amount of pension is determined by a simple formula. Here are some examples of its working. An employee retiring at 50 gets $504 a year. If he retires at 60, he gets $630 a year; at 70, $840 a year; at 80, $1,260 a year.

One employee known to me has just retired with a pension of $700.

What would his pension have amounted to had he waited another year?

24 DOWNING STREET

A knot of spectators in Downing Street was watching members of the Cabinet as they arrived for a critical meeting.

"Who's that?" I asked my neighbor, as a silk-hatted figure, carrying a rolled umbrella, rang the bell at No. 10. "Is it the Minister of Morale?"

"Yes," he said.

"Quite right," said a second spectator. "The Minister of Morale it is. Looks grim, doesn't he?"

The first of these speakers makes a point of telling the truth three times out of four. The second tells the truth four times out of five.

What is the probability that the gentleman in question was in fact the Minister of Morale?

25 CRANKSHAFT

At noon, Crankshaft, who is in training for a cycle race, left the Three Tuns at Shipton to ride to the Stork at Rosehill and back again. It is 26 miles each way. Crankshaft did the double journey without stopping, and maintained a uniform speed throughout.

Some time later Gearbox, trying out his new car, left the Stork and drove—also maintaining a uniform speed—to the Three Tuns and back again. Gearbox passed Crankshaft, on the latter's outward journey, $7\frac{1}{2}$ miles from the Stork, and passed him again, on his return journey, $5\frac{1}{2}$ miles from the Stork. Gearbox finished the double journey at 3:20 P.M.

What time was it when Crankshaft was back at the Three Tuns?

26 MARBLES

I have two little bags, of which the contents are identical. Each has in it four blue marbles, four red ones, and four yellow ones.

I close my eyes and remove from Bag No. 1 enough marbles (but only just enough) to ensure that my selection includes two marbles at least of any one color, and one marble at least of either of the other colors. These marbles I transfer to Bag No. 2.

Now (again closing my eyes) I transfer from Bag No. 2 to Bag No. 1 enough marbles to ensure that, in Bag No. 1, there will be at least three marbles of each of the three colors.

How many marbles will be left in Bag No. 2?

27 MUCH SPENDING

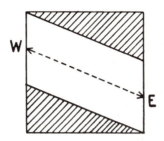

There has been some criticism by the rate payers of the new housing estate at Much Spending. Its site is (as shown) a perfect square, each side being three-quarters of a mile.

Houses are only to be built on the shaded (triangular) areas; these two triangles are of identical proportions. The intervening space, down the center of which an avenue runs from West to East, is to be occupied by a communal garden.

It is the devotion of so much space to this garden that has provoked adverse criticism. It occupies seven-twelfths of the area of the estate.

What is the length of the central avenue?

28 ALGEBRAICA

The republic of Algebraica lies so far behind the Iron Curtain that very few of us have heard of it. This republic, anxious to flout "capitalistic conventions and superstitions," has introduced a scale of notation containing three extra digits—X, Y, and Z—which are interpolated, very confusingly, in the ordinary denary scale. Thus we have:

Our numbers:	1	2	3	4	5	6	7	8	9	10	11	12	13
Their numbers:	1	2	3	X	4	5	Y	6	7	8	Z	9	10

Our 20 is represented by 1Y; our 100 by 77; and so on.

What (in the notation of Algebraica) is the square of their number 1X?

29 CLOWNE WAS UNLUCKY

Mr. Clowne, the impresario, and four of his friends crossed the Atlantic together. On each of five successive days they ran a private sweep on the length of the day's run. Mr. Aphis was successful twice; Mr. Birdseed, Colonel Dimwits, and Mr. Edelweiss each won once. Clowne (as you will have gathered) had no luck.

"What were the odds," Mr. Clowne asked me, "against my drawing a blank every day? Two to one—is that right?"

What is the answer?

30 LITTLE MATING

Twenty-four players competed in the recent Chess Tournament at Little Mating.

The committee divided them into two sections. In each section each player played one game against every other competitor.

There were 69 more games in Section B than in Section A.

Mr. Gambit, unbeaten in Section A, scored $5\frac{1}{2}$ points. (A win = 1 point; a draw = $\frac{1}{2}$ point.)

How many of Gambit's games were drawn?

31 TETRAHEDRAL

Each of the four faces of a regular tetrahedron is divided as shown into four equilateral triangles. I am given four

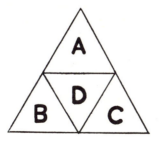

colors with which to paint this solid, and I must paint it in
accordance with the following rules:

(1) Only two colors to be used on each face.

(2) Four triangles to be painted in each of the four colors.

(3) No two triangles having one side in common to be
painted the same color.

**In how many different ways can I paint my
tetrahedron?**

32 PEDAL AND HOOFIT

It is 16¼ miles from the Black Bull to the Anchor. At noon
the other day Hoofit left the Black Bull to walk to the Anchor,
and Pedal left the Anchor to walk to the Black Bull. Each,
on arrival, immediately mounted a bicycle and cycled home.
Both reached home as the clock was striking six.

Hoofit is the better pedestrian, but on his bicycle he travels
only half as fast again as on foot. Pedal, on the other hand,
cycles just three times as fast as he walks.

**How far from the Anchor were the two cyclists when
they passed one another on the journey home?**

33 MRS. INKPEN

Mr. Pisistratus Patriarch lived up to his somewhat unusual
name. He had nine children, and no fewer than 31 grand-
children.

In his will he left an exact number of dollars to each grand-child. Each girl was to get $7 more than each boy. All 31 grandchildren were alive when Patriarch died, and their legacies totaled $470.

Of this amount, $74 went to Mrs. Inkpen's children (she was Patriarch's eldest daughter).

How many daughters had Mrs. Inkpen?

34 THE TOOTLES

Mr. and Mrs. Tootle were captured by Chinese brigands.

"I propose to hold you to ransom," said Na Poo, the brigand chief. "But I'll give you a sporting chance of getting away free.

"Here's a circular table. There are six chairs which, in clockwise order, are labeled A B C D E F. You can seat yourselves in any two chairs. You may then throw two ordinary dice—one each. I shall take the combined total, which may be any number from 2 to 12. Then, starting with A, I shall count out two chairs, taking the number thrown, on the 'eeni-meeni-mo' principle.

"Thus, if you throw 2, I count out B and D. If you throw 11, I count out E and F. Is that clear?"

"Perfectly," said Tootle.

"The occupant of a chair counted out will not be held to ransom," said Na Poo.

The Tootles chose chairs B and F. Could they have done better?

35 HELPUSELPH

A settler in the island of Helpuselph applied to the Governor for some land. "How much would you like?" asked the Governor.

"About 100 square miles."

"Okay," said the Governor. "You may choose a rectangular parcel of land in the township of Little Rainfall. Its dimensions must be such that, if one side of the rectangle were 5 miles longer, and the other 4 miles longer, the area of the rectangle would be twice as great; and its perimeter must be exactly 46 miles."

The applicant duly selected and fenced his land in accordance with these conditions. But he got away with six square miles more than the Governor had anticipated.

What was the area of the selected rectangle?

36 DROPPED CARD

Playing bridge the other day, I noticed that a player at another table held the Ace, King, Queen, and Knave of Hearts. I saw none of his other cards. During the bidding, which he opened (not unnaturally) with "One Heart," one of his cards fluttered, face downwards, to the floor.

What is the chance that the card in question was a Heart?

37 HIGHTONE

That select establishment, Hightone Academy, plays its own particular brand of football. It is easier to score goals in this game than in ordinary soccer, and a game must go on until eleven goals in all have been scored. Hence there can never be a drawn game.

The Challenge Cup is competed for by four Houses. Each plays one game against each of the others. Last year Bacon put up the best showing, the final results being: Bacon, 2 wins, goals scored 23–10; Emerson, 2 wins, goals scored 18–15; Ruskin, 1 win, goals scored 13–20; Carlyle, 1 win, goals scored 12–21.

No two matches produced the same result. Ruskin beat
Emerson by 8 goals to 3.

How did Carlyle fare against Bacon?

38 POLYCHROME

Greene and Scarlett (of the Polychrome Harriers) are in
training for some long-distance races. The other day they
ran, in opposite directions, around a four-mile track; they
started at the same time, and finished at the same time.
Greene maintained a uniform pace throughout. Scarlett
maintained a uniform pace for two miles; then, for the second
two miles, increased his speed by no less than four miles an
hour.

When the runners passed one another, Greene had covered
2 miles and 320 yards.

**How long did the two athletes take to complete the
circuit of the course?**

39 CHESS

Representatives of Doomshire and Gloomshire met to play
chess over 100 boards. Both men and women players took
part, each county producing more than 50 men players.

More women played for Gloomshire than for Doomshire.

It was arranged that, so far as possible, men should be
matched against men and women against women. This
meant that there were only mixed matches (men versus
women) to the extent to which the number of men players
representing Doomshire exceeded the number of men representing Gloomshire.

On the boards where men were matched against men,
Doomshire won three matches out of five. On the boards
where women were matched against women, Gloomshire
won two matches out of three. And on the boards where

Doomshire's surplus men met Gloomshire's surplus women, the Doomshire players won two-thirds of the matches. No game ended in a draw. The result was a very narrow win for Doomshire by 51 games to 49.

How many men played for Doomshire?

40 SNATCH

"What's happening in the card room?" I said to a friend at the club.

"Puffin and Snatch are playing bridge against Hobo and Hardlines. Puffin has just dealt four interesting hands."

"Tell me about them."

My friend looked quizzical. "Puffin and Snatch," he said, "hold most of the red cards. Their red cards—let me think, now—exceed the total of their opponents' red cards by the number of black cards in Hobo's hand. Puffin holds twice as many black cards as Hobo holds red ones."

"I don't see what I can deduce from that," I said.

"No? Not even if I tell you that Hobo has both the red Aces? He might easily make six No Trumps."

How many black cards does Snatch hold?

41 PRIVATE ENTER-PRISE

Mr. Egbert Enterprise (who had been known in the Army as Private Enterprise) bought a stock of ex-Army bicycles. It took him just nine months to dispose of them.

"Sales resistance" was at first considerable. But each month, after the first, Enterprise reduced the price of each bicycle by $1. As a result, he sold each month (after the first) four more bicycles than he had sold the month before. His selling price was always an exact number of dollars.

His gross receipts were $3,153. He had paid $7 per bicycle.

In which month did he make the largest profit?

42 DIVIDING THE PACK

A pack of cards is divided by cutting into two unequal portions.

If a card is drawn at random from Portion A, the odds are 2 to 1 against its being a red card.

A red card is next transferred from Portion B to Portion A. Now the odds are 2 to 1 against a card drawn at random from Portion B being black.

How was the pack originally divided?

43 TROGLODYTES

Extract from a report of the Games Sub-Committee of the Ancient Order of Troglodytes:

"We recommend the purchase of 2 billiard tables and 12 dartboards. A poll of our members shows that 17 per cent would like to play billiards, 22 per cent snooker, and 28 per cent darts. These percentages include 137 members who have said they would like to play all three games. No one expressed a wish to play two."

There was some grumbling by 1,561 Troglodytes who did not want to play any of the three games.

How many Troglodytes are there in all?

44 SPEEDWELL

Jack Speedwell, the road cyclist, left Cloudburst at 9:00 A.M. yesterday and rode nonstop to Grizzle, 27 miles away. He passed several pubs, including first the Red Lion and then the Porpoise, which is six miles from the Red Lion, and nearer to Grizzle than to Cloudburst.

From Cloudburst to the Porpoise, Jack's average speed in miles per hour equaled the distance in miles between the Red Lion and Grizzle. From the Porpoise onwards, Jack's speed in

miles per hour equaled the distance in miles between Cloudburst and the Porpoise. He reached Grizzle at 10:42 A.M.
How far is the Red Lion from Cloudburst?

45 SEVEN DIGITS

"How's your mental arithmetic, Cicely?" said Dr. Dingo to his daughter.

"Not too hot, I'm afraid."

"Then have a go, as they say on the radio." Dingo took out his watch. "I'll give you three minutes for this one."

"Go ahead," said Cicely. "I trust I get the usual dollar?"

"Fifty cents is enough for this one. . . . I'm thinking of a number. I've squared it. I've squared the square. And I've multiplied the second square by the original number. So I now have a number of seven digits."

"Big of you," said Cicely. "Where do I come in?"

"You tell me the number I originally thought of. I'll give you a clue, if you'd like one. The last digit of my seven-digit number is a 7."

How quickly can you discover the number that Dingo first thought of?

46 GLOOMSHIRE

Four teams competed last season for the soccer championship of Gloomshire: Much Moaning; Jitters; Gruntle; and Deadalive. Each met each of the others once, on three successive Saturdays.

On the basis of two points for a win and one point for a draw, each team scored three points. Much Moaning won the championship on goal average. The goals "for and against" were: Much Moaning 5–1; Jitters 5–5; Gruntle 3–3; Deadalive 3–7.

On the first Saturday Gruntle beat Jitters by the odd goal in five. Jitters did better the following week, when they drew 2–2 against Deadalive.

What were the precise results of the two matches played on the third Saturday?

47 NO LUCK AT ALL

"Any luck at bridge last night?" I asked my old friend Reggie Reckless.

"None," he replied. "It was a very expensive evening. I played four rubbers and lost them all."

"High stakes?"

"Not to begin with. But after the first rubber I trebled them; after the second rubber I trebled them again; and after the third rubber I trebled them once more for the fourth. Then I packed up, having lost $596."

"Were they large rubbers?"

"Not very," said Reckless. "Each rubber after the first was actually 100 points less than the preceding one. We played at so much a hundred, of course."

How large was the first rubber?

48 FLOATING VOTE

Young Valentine Vote was floating down the river on a raft when, half a mile lower down, his brother Victor took to the water in a skiff. Victor pulled downstream at the best pace of which he was capable (which in still water is ten times that of the current); then turned around and pulled back again; and arrived at his starting point just as Valentine drifted by.

What distance had Victor covered before he turned his boat around?

49 PUTWELL WAS NETTLED

"I seem to be all over you today," said Foozleham to Putwell, as they stood on the twelfth tee.

Putwell was nettled. "Here's a bet for you then," he said. "We have still seven holes to play. I'll pay you money on the following basis: $1 for a single hole won by you; $3 for two consecutive holes; $6 (that's 1 plus 2 plus 3) for three consecutive holes; $10 for four consecutive holes, and so on. If you win all seven you get $28. I'll pay you on that basis, if you'll pay me on the same basis for each hole or series of holes that you don't succeed in winning.''

"Done," said Foozleham. "It's a cinch."

He won four holes, and Putwell won three, but no money changed hands.

How did the account stand at the end of 17 holes?

50 WHOPPIT

I wonder if cricket statisticians have noticed some interesting facts about those doughty batsmen, Bludgeon, Whoppit, and Thrust—all of whom retired last season after playing for many years for Battleshire.

Bludgeon's gross total, over the years, was exactly the same as Thrust's. Bludgeon's annual total was, on the average, 300 more than Whoppit's; but Whoppit's was, on the average, 300 more than Thrust's. On the other hand, Thrust had been in county cricket three years longer than Whoppit and six years longer than Bludgeon.

How does the total number of runs amassed by Whoppit compare with the total amassed by Bludgeon or by Thrust?

51 RABBITT

My friend Rabbitt (who lives in the country) caught an earlier train home than usual yesterday. His car normally meets him at the station. But yesterday he set out on foot to meet it, and so reached home 12 minutes earlier than if he had waited at the station. The car travels at a uniform speed which is five times Rabbitt's speed on foot.

Rabbitt reached his home just as it was striking six. **At what time would he have reached home if his car, forewarned of his change of plan, had met him at the station?**

52 LARRY

"Just ten past five," said my cousin Larry, as he landed at the boathouse. "Not bad going. I've done three miles upstream, and three miles back again. Rowing at a steady pace, both ways."

"Current's flowing strongly, isn't it?" asked someone.

"I'll say it is," said Larry. "It took me just half the time to row downstream as to row upstream. What do you know about that?"

Assuming that Larry would have progressed, in still water, at a steady five miles per hour, **what was the speed of the current?**

53 MRS. ANTROBUS

"I always answer my daughter's questions truthfully," said Mrs. Antrobus, smugly. "Fibs, however well-intentioned, must react unfavorably on a child's upbringing."

"What would you say if she asked you how old you are?" asked someone. "Most mothers, I've noticed, are inclined to be vague about that."

"She did ask me that, only yesterday," said Mrs. Antrobus. "She also asked me what my own mother's age is. I gave her a truthful answer, though not an intelligible one. 'Darling,' I said, 'if you squared your granny's age (in years) and also squared my age (in years), and subtracted the second square from the first, the difference would be 2,720.'"

"And what did darling say to that?"

"She said: 'Ooh, you are old, mummy, aren't you? Much older than Fido.'"

How old is Mrs. Antrobus?

54 DEEPDENE

"Our last interhouse soccer competition offers the material for an intelligence test," writes the headmaster, Dr. Livy, from Deepdene. "We have, as no doubt you know, four Houses—Ovid, Vergil, Horace, and Catullus. Each plays one match against each of the others. Last season no two matches produced identical scores, and this was the final competition table:

	W	D	L	GOALS F	A
Vergil	2	0	1	5	1
Catullus	2	0	1	3	5
Horace	1	0	2	5	6
Ovid	1	0	2	4	5

"A little exercise based on these figures would be much appreciated by my upper forms."

The figures are certainly interesting, and here is a "little exercise" which suggests itself: **What was the score in the match between Vergil and Ovid?**

55 PEWTER

"My Uncle James is an old man," said Snoggins. "He's beginning to disembarrass himself of his worldly goods. He assembled all his nephews, and distributed among us his famous collection of pewter pots. There were once nearly 100 of them, but quite a few were stolen. I came away with sixteen of them; so did my brother Luke. But all of us got at least one pot."

"But some were more favored than others, I take it?"

"Not really," said Snoggins. "The old guy has a nice sense of justice. He gave away money as well as pots; we weren't too proud to accept it. Each of us received $5 in respect of each pot that didn't come to him; less $10 in respect of each pot that did. Ingenious, don't you think so? $510 in all was distributed, as well as all the pots."

How much did Snoggins receive in cash?

56 SCHOLARSHIP

Five candidates competed for an Entrance Scholarship at Zacharias. The basis of marking was devised by Dr. Tombworthy. The candidate who takes first place in a subject gets one mark; the second candidate, two marks; and so on. There were five subjects in all, the candidate with the lowest aggregate (in this case 12 marks) securing the scholarship.

Allbrane, the winner, was two marks up on Bookish. The other candidates, in order of merit, were Cramwell, Drudge, and Gigadibs. Each of the five took first place in one subject. Allbrane was first in Greek; Bookish, first in Latin and third both in French and in English; Cramwell, first in French, second in Mathematics, third in Greek; Gigadibs is no linguist, but he took first place in Mathematics and second place in English.

What was Drudge's mark in each of the five subjects?

57 FIVE RUBBERS

"Bad show at the club last night," said old Tom Tenace. "Played five rubbers at highish stakes and lost them all. Set me back $1,415."

"What were the stakes?"

"Progressive stakes, as we call 'em," said Tenace. "I forget what we started with—so many dollars a hundred. Each rubber after the first went up a dollar a hundred on the one before."

"So each rubber after the first cost you more than the previous one?"

"That doesn't follow," snapped Tenace. "Actually—this is rather interesting—each rubber after the first worked out at 300 points less than the previous one."

I had myself partnered Tenace in his first two rubbers; the other three I played against him.

How much did I lose or win?

58 ELGIN'S MARBLES

"I have 18 marbles in these two little bags," said Mr. Elgin. "Some of them are red; the others are blue. If one marble is drawn at random from one bag, and one from the other, the odds against their both being red are 11 to 5. What do you suppose are the odds against their both being blue?"

"I shouldn't think you can say with certainty," said someone. "How many are there in each bag?"

"That," said Mr. Elgin, "I'm not telling you. But there's only one answer to my question, and you should be able to discover it."

What is the answer?

59 MILD GAMBLE

"Would either of you ladies like a mild gamble?" asked Sharpwits. He was waiting, with Miss Dimple and Mrs. Large, for a fourth at bridge.

"What sort of a bet?" asked Miss Dimple.

Sharpwits took, from a pack of cards, four of each suit; shuffled his 16 cards; and presented them to Mrs. Large to cut. "I'll deal you three cards each," he said. "And I'll lay odds, either against the three cards which either of you gets being all cards of the same suit, or (alternatively) against their being all three different suits."

"What odds do you offer?" said Mrs. Large.

"Thirty to one," said Sharpwits, "against all three cards being of the same suit. Evens against all three being cards of different suits."

Which is the more advantageous offer from the punter's point of view?

60 BRACELETS

You have a number of beads of the same shape and size but of three distinctive colors. Let us assume that the colors are red, green, and yellow.

Bracelets are to be made, each consisting of five beads. Of the five beads used for each bracelet, two are to be of one color; two of a second color; one of a third color.

How many distinguishable bracelets can be threaded? Don't forget that a bracelet may look different when viewed from the other side.

61 SIX BOWLER HATS

Tom, Dick, and Harry are brothers. They wear almost identical bowler hats. If each takes a hat at random from the hatstand, there are two chances in six (i.e., one chance in three) that none will have taken his own hat. For the three hats can be taken in six different ways, and two of these give each of the three men a hat other than his own.

Here is a more interesting puzzle. Each of the three brothers has two bowler hats on the hatstand. Each takes a hat at random. **What is now the chance that none of the three has taken one of his own hats?**

62 DARTS

Each of five public houses in the same town has a highly proficient darts team. In a contest organized last year each house played one match against each of the others. The matches took place on five successive Saturdays: two matches each Saturday, while one team had a bye. On the first Saturday the Feathers played the Crown. On the second Saturday the

Unicorn played the Bull; the Unicorn, who won this match, triumphed the following week against the Sceptre. On the fourth Saturday the Bull had the bye.

Who had the bye on the fifth Saturday?

63 CHESTNUT AVENUE

"What a long street Chestnut Avenue is," said Jones, a visitor to the thriving city of Middlesbrum.

"Yes, isn't it?" said his host. "It's our principal residential center." He took out his pocketbook. "No fewer than eighteen of our friends live in Chestnut Avenue. I'll copy their numbers down for you."

He proceeded to write down the following:

1, 3, 13, 16, 21, 27, 28, 39, 52, 63, 70, 78, 156, 175, 189, 208, 243, 256.

"What do I want those numbers for?" asked Jones. "Do I have to call on all your friends?"

"Far from it, my dear fellow. The numbers happen to have a fascination of their own. Try and arrange them in six sets of three—in such a way that all six sets have two obvious properties in common."

This simple exercise defeated Jones. **Can you succeed where he failed?**

64 THE LIARS

Lizard, Stoat, and Flittermouse are congenital liars: they lie systematically and on principle. Stoat is the worst liar: out of every three statements which he makes, two (precisely) are lies. Flittermouse, similarly, lies three times out of five. Lizard is comparatively truthful; of the statements which he makes, only 50 per cent are lies.

These three, together with their friend Puddock (who is also a liar, though this fact is irrelevant) adjourned the other night

to the "local." Here the four of them took part in a knockout
tournament at Snooker. There were no other competitors.
When I asked them who won the tournament, Lizard said:
"I did of course." Flittermouse said: "No, Stoat won"; and
Stoat said: "My friend Flitters flatters me; he won the tourna-
ment himself."

**How does the chance that Flittermouse won the
tournament compare with the chance that it was
actually won by Puddock?**

65 MR. STOUT AND MR. PORTER

Mr. Stout and Mr. Porter, who had just received their weekly
pay packets, went into the Three Tuns for a quick one.

"Toss you who pays," said Stout.

"Okay," said Porter. He took five coins from his pocket.
"Let's each throw all these coins on the counter. Then whoever
throws more 'heads' wins, and the other bloke pays for the
drinks."

"Suppose we throw the same number of heads?" said Stout.
"Do we toss again, or what?"

"Pooh," said Porter. "That's not likely to happen. But, if by
any chance it does, I'll pay."

**What, in fact, are the chances that Mr. Porter pays for
the drinks?**

66 SNOOKER

Messrs. Green, Pink, White, Blue, and Brown entered for a
snooker contest. Each was to play one frame against each of
the others. There was little difference in skill between any two
of them, which made it probable that no player would win
four frames.

"So here's an idea," said Colonel Black, chairman of the committee. "Let each player score one point for every frame he wins, plus half a point for every frame won by each player whom he defeats. The player who, on this basis, scores most points will be the winner of the contest."

The Colonel's plan was adopted. Pink won the contest; Green was the runner-up. Collectively, the four players scored an aggregate of 18 points.

How many frames did Green win?

67 DIMONT WAS UNLUCKY

"Did you play bridge late last night, Harz?" asked his friend, Tenace.

"Dear me, no," said Harz. "We only had three rubbers. There were just the four of us: Dimont, Clubb, Spaight, and me. We pivoted, as they call it: a different partnership each time."

"Any luck?"

"I won twenty-five dollars. Dimont was the unlucky one: he lost all three rubbers."

"High stakes?"

"So-so," said Harz. "So many dollars a hundred the first rubber; stakes doubled for the second rubber; doubled again for the third. The second rubber—fortunately for Dimont— totaled 300 points less than the first; the third rubber came to 300 points less than the second. I was Dimont's partner in the last one, when the stupid fellow missed a sitting slam."

Tenace made a rapid mental calculation. "Bless my soul," he said. "Dimont must have lost $2,000 or so."

"Don't be absurd," said Harz. "The stakes weren't anything like as high as that."

How much, in fact, did Dimont lose?

68 POSTAGE STAMPS

"I have 1,000 stamps to give away," ran the notice which Mr. Squeers, a retired schoolmaster, put up in the village hall. "If any young philatelists like to call on me, I'll see what I can do for them. I shall be at my house daily, between 10 A.M. and noon, from Monday next to Friday inclusive."

On each of the five appointed days claimants to stamps appeared. Business was not very brisk at first; but each day, after the first, two more turned up than had appeared the day before. Squeers adjusted his distribution accordingly. Each applicant, after Monday, received ten fewer stamps than each applicant had received the day before.

Squeers' stock was exhausted on Friday, when he had fewer than 100 stamps to give away.

How many collectors in all took advantage of his offer?

69 TWENTY-ONE RUBBERS

Messrs. North, South, East, and West spent seven days together on a fishing holiday. On each of the seven evenings they played three rubbers at bridge, when each player was partnered for one rubber by each of the others. The stakes were one dollar per hundred points.

Mr. North, who is interested in card statistics, kept a careful record of these 21 rubbers. Money had changed hands after every rubber, and North had won them all. He had, moreover, won the same amount every night. This was somewhat odd, as on no two of the seven nights was the total sum. won by Mr. North made up of the same three components. And also (North noted) the total which he won each night was the smallest possible, consistent with these data.

Mr. East was the most successful—or should one say the least unsuccessful?—of the other three players. Neither South nor West, on any one night, won more or lost less than he. Mr. South, on the other hand, never won more, or lost less, than either of the others.

How much did Mr. North win during the week, and how did each of the other three players fare?

70 RUY LOPEZ

I was recently discussing with a fellow chess enthusiast the continuing popularity of the classical opening known as the Ruy Lopez. He drew my attention to an Inter-Schools Tournament with which he had lately been concerned.

"It's run on rather odd lines," he said. "Each competing school enters a team of from three to eight players. Each member of the team plays one game against every member of each of the other teams. This may mean, of course, that some players get a good many more games than do other players; so the place of each team in the final table is determined on a percentage basis. A rough-and-ready system, but it works reasonably well.

"This last year, three schools competed. There were 14 players altogether, representing the three schools. And the Ruy Lopez was the opening adopted in just one-third of the games played.

"Pawnborough School, who secured the championship, either opened with the Ruy Lopez, or defended it, in one-fifth of their games. In the case of the Two Bishops School, the proportion of Ruy Lopez games was one-quarter."

In how many Ruy Lopez games was the third school— Castleton—concerned?

71 THERE AND BACK

"Your brother, like you, has a new car, I see," I remarked to Colonel Crankshaft. "How does his model compare with yours?"

"Not much in it, we think," said the Colonel. "We tried out a little experiment last week. Starting from my place, we both drove to the Market Cross at Hightown, which—as I dare say you know—is just 40 miles away. On arrival, we turned round and drove back again. And—believe it or not—we were back at my place at practically the same moment."

"You made a race of it, did you?"

"Not at all. That would have been very naughty. No; we just cruised along. I drove to Hightown at a uniform speed and back at a uniform speed; so did Bob, my brother. But I came back at just double my outward speed—it's mostly downhill, you know—while Bob's return speed was only 25 per cent greater than the speed at which he drove to Hightown. I passed him on my outward journey just opposite the Blue Griffin."

How far is the Blue Griffin from the Market Cross at Hightown?

72 ISLE OF MUGS

The Isle of Mugs, as everyone knows, encourages motor racing on its roads. There is a recognized circuit used by road racers which begins and ends at the White Swan Inn. One day recently, Spaghetti and Gnocchi left the White Swan in their racing cars as the clock was striking 12. Each proceeded to go twice round the circuit: Spaghetti in a clockwise and Gnocchi in an anticlockwise direction.

Gnocchi, who throughout his first circuit had maintained a uniform speed, increased it for his second circuit by 30 m.p.h. Spaghetti maintained throughout both circuits a uniform speed, which was 12 m.p.h. faster than was Gnocchi's speed during his first circuit. The two drivers completed their second circuit at precisely the same moment.

They passed one another for the first time opposite Mugtown Village Hall. They passed one another for the third time opposite the Red Lion. The Red Lion and the Village Hall are exactly a mile from one another.

How far were the two drivers from the White Swan when they passed one another for the second time?

73 FUGITIVE FOUR

Mr. Black and Mr. White were filling up football pool coupons. They were entering for a simple gamble called the Fugitive Four. In this competition one has to try to forecast the results of four games: the result can be either a home win, an away win, or a draw.

"I shall mark my results at random," said Black. "That is to say, I shall throw four dice, one after the other. If my first die comes up 1 or 2, I shall put down a home win for the first match; if 3 or 4, I shall forecast an away win; if 5 or 6, a draw. And similarly with the next three matches."

"A good idea," said White. "Lend me your dice when you have done with them, and I'll do just the same thing."

Before throwing their dice, they spent some time arguing which would be most likely: that their "forecasts," determined at random, would turn out to be the same in respect of (*a*) all four matches; (*b*) three matches; (*c*) two matches; (*d*) one match; (*e*) none of the four matches. Mr. Black said it was "long odds" that their method would not produce the same forecast in respect of any match; Mr. White thought that "most probably" they would produce the same forecast in respect of just one of the four matches.

Which of them—if either—is right?

74 FIREWORKS

Grandfather has always enjoyed fireworks, and this year, having just celebrated his birthday, he bought as many rockets as he is years old to mark the occasion.

But half of them got damp, the children borrowed a third of the good ones for a private celebration, and the sticks were missing from 21 others.

"Never mind," said Grandfather, "it just means that there'll be one for every ten years, instead of one for every year."

It was a poor show, though. We'll have a better one when Grandfather is 100. **When will that be?**

75 MUNICIPAL PARK

A municipal park has recently been opened in the ancient city of Danechester. The park is rectangular in shape. When I asked what its dimensions were, I was given two somewhat odd items of information. The first: That the diagonals of the park, plus its longer sides, were together equal to seven times one of the shorter sides. The second: That the length of one diagonal exceeded that of one of the shorter sides by just 250 yards.

What is the area of the park?

76 GREENLEAF

"You want a train for Greenleaf?" said my host. "I haven't a timetable, unfortunately. All I can tell you is that, if you like to take a chance, the odds are 7 to 3 in favor of the next train from the down platform being a Greenleaf train."

I made further inquiries and found that the down trains ran alternately, to Greenleaf and to Blackthorn. There was a train every half-hour to Greenleaf, and a train every half-hour to Blackthorn. When I got to the station I found that the train which had last left the down platform was the 9:22 for Blackthorn.

At what time did my train leave for Greenleaf?

77 WRECK OF THE *HESPERUS*

When an emigrant ship, bearing the ill-fated name *Hesperus*, was wrecked off Turtle Island, the authorities voted the sum of $1,000 as a donation to the survivors. This sum was divided among the men and women who were eligible, and their 20 children, in accordance with an agreed scale. Each child received so many dollars exactly; each woman received six times as much as a child; each man, $5 more than each woman. There were just twice as many men as women.

The *Hesperus* carried 118 passengers. **How many of these survived?**

78 INSECT LEAGUE

Last season four soccer teams organized themselves as a miniature "league." Each played one game against each of the others. No two games produced the same score, and each of the four teams won one game, drew one, and lost one. So the destination of the "league championship" turned upon goal average. The Hornets had scored 4 goals against 2; the Butterflies, 2 goals against 2; the Wasps, 3 goals against 4; the Dragonflies, 2 goals against 3.

What was the score of the game between the Dragonflies and the Wasps?

79 BUTTER-FINGERS

At the Club the other night old Professor Buffington—"Butter-Fingers" they call him—dropped some of his cards on the floor as he was sorting his hand in a bridge game. I picked them up for him and noticed that all of them were Clubs. He thanked me, and proceeded to sort his hand anew. Just as he was about to play to the first trick he dropped another card, this time face downwards. I amused myself by calculating what the odds were against its being a Club, and decided that they were 9 to 7.

How many Clubs did Butter-Fingers drop originally?

80 FAMILY SWEEP-STAKE

The Gamblers and the Plungers—two neighboring families who are on very friendly terms—recently organized a "holiday sweepstake" to amuse their numerous offspring. One ticket was allotted to each child, and there were two handsome prizes.

Before the draw, Mr. Gambler and Colonel Plunger were discussing its relevant mathematics. "It's just as likely," said Gambler, "that the prizes will go to two of my daughters as that they'll go to two of yours. That's obvious. On the other hand, the odds against their going to two of my sons are four to one, while the odds against their going to two of your lads are longer: six to one. That's right, isn't it?"

"Perfectly right," said the Colonel.

Actually, the first prize went to one of Gambler's sons, and the other to one of his daughters.

What, before the draw, were the odds against this?

81 GRANDPA WAS EIGHTY-NINE

"Grandpa was eighty-nine yesterday," said Cissie to her younger brother James.

"You're telling me," said James. "I've been going into this question of ages. Grandpa's being 89 just completes my set of coincidences."

"Coincidences?"

"Age coincidences," said James. "If you multiply Grandpa's

age by itself; multiply Grandma's age by itself; and take the difference—if you do that, you find the difference is the same as the difference between the square of Aunt Julia's age and the square of Uncle Pongo's age."

"Who d'you call Uncle Pongo?"

"Aunt Julia's husband."

"Very odd," said Cissie.

"But that's nothing like the lot," said James, pulling out a grubby notebook. "You get the same difference—the same, mark you—between the squares of Mr. Brown's age and Mrs. Brown's; between the squares of Uncle Guy's age and Aunt Clara's; between the squares of Uncle Will's age and Aunt Mary's; between the square of Dad's age and the square of yours; between the square of Mum's age and the square of mine; and between the square of Aunt Fritillary's age and the square of Baby's."

"I don't believe it," said Cissie.

"'Strue, all the same," said James.

How old is this precocious youth?

82 RATHER SILLY

"What were you doing this morning?" I asked Betty.

"That thing called the harmonic mean."

"Does it interest you?"

"Well, I thought it rather silly."

"Why silly?"

"Take this example," said Betty. "I was working out the harmonic mean of our ages: Baby's, Belinda's, and mine. (I'm the eldest.) Then I worked out what their harmonic mean would be in five years' time. And I found that my second mean exceeded my first, not—as I'd expected it to do—by 5 years, but by 5 years and 95/143 of a year! That's what I call silly."

"Dear me," said I.

How old is Betty?

83 NEWS FROM CRAB'S BAY

It is 52 miles by road from Crab's Bay to Prawnacre.

At 10 A.M. yesterday Peter Pedal, who is in training for a cycle race, left Crab's Bay and rode at a uniform pace—and without stopping—to Prawnacre and back again.

Some time later, Mr. Magneto, trying out a new car, left Prawnacre and drove at a uniform speed to Crab's Bay and back again.

Magneto passed Pedal (on his outward journey) 15 miles from Prawnacre. He passed him again, on his return journey, 11 miles from Prawnacre.

Magneto was back at Prawnacre at 4:40 P.M.

What time was it when Pedal was back at Crab's Bay?

84 ESSAYS

"Six of my children," said Mrs. McScribble, "are taking part in an essay contest. There are so many prizes for boys; so many for girls. And it is possible—though only just possible —for my family to carry off all the prizes."

"How many competitors are there?" I asked.

"Twenty. Over half of them are girls."

"Ha!" I said. "And you're about to ask me to deduce how many of your family are girls. Having first—presumably—given me the odds against your family taking. all the prizes."

"Something like that," said Mrs. McScribble. "Of course, the odds can only be calculated on the assumption that all the boys competing are equally likely to get a prize; and similarly with the girls. I've worked these odds out. And I find that, if one of the girls competing withdrew, and a boy entered in her place, the odds against my family's securing all six prizes would remain exactly the same."

(1) How many boys, and how many girls, are competing?

(2) How many of Mrs. McScribble's children are girls?

85 NARROW TRIUMPH FOR CLOWNCHESTER

"Did we win?" said Professor Probe to his colleague, Dr. Bunn. He was referring to Clownchester's annual Athletics Contest against the Crab's Bay Polytechnic.

"Yes," said Bunn. "We just pulled it off, Professor. By 25 points to 23."

"How did my boy, Peter, get on?"

"Covered himself with glory," said Bunn. "He was first in two events, and second in a third."

I should explain that, in each event, points are awarded for first, second, and third places. Clownchester had taken four firsts, two seconds, and one third.

How many of Clownchester's points were scored by Peter Probe?

86 ALLOTMENTS

Adam, Barnabas, Charlie, and Dave all cultivate allotments at Little Holdings. All four allotments are rectangular; each side of each allotment is an exact number of feet; and the diagonal of each allotment is—strange to say—exactly 221 feet.

They were recently comparing the areas of their several "cabbage patches." Adam's exceeds Charlie's by 3,660 square feet; while Barnabas has no fewer than 12,720 square feet more than Dave to cultivate.

By how much does the area of Adam's allotment exceed the area of Dave's?

87 THE PROFESSOR'S DAUGHTER

"Let me see," said the Colonel to the Professor; "your youngest daughter is—how old? I haven't seen her for at least three years."

"I'll give you three guesses," said the Professor. "Here are some data for you. If my daughters' ages (in years) were multiplied together, the product would equal the number of cents in twelve dollars. And if their ages were added together their sum would fall ten short of my wife's age."

"Too tenuous," said the Colonel. "How old, if you please, is your wife?"

The Professor obligingly supplied the information.

"H'm," said the Colonel. "I never make a mistake in arithmetic." (This we can take to be true.) "Your youngest daughter's age is the same as my niece's."

"Wrong," said the Professor.

"Then she's the same age as my nephew."

"Wrong again," grinned the Professor.

How old is the Professor's wife?

88 JONES COLLECTED ONE DOLLAR

"While we're waiting for the others," said Jones to Brown, "let's do something with these dice."

"Okay," said Brown. "What do you suggest?"

"We'll each throw two dice," said Jones, "and multiply together the numbers which turn up. For example, if you throw a 6 and a 3, your product is 18. Then the one with the lower product pays the other $1 per point on the difference. One's maximum gain would thus be $35."

"Suits me," said Brown. "If we get the same product, we both throw again?"

"That's right."

Jones threw first. His throw produced a 4 and a 3.

"Ha," he said. "That's not too bad. There are 19 chances in 36 that you get a lower product, and only 13 chances that you get a higher one. . . . Like to pay me a dollar to call the whole thing off?"

"I think I'd better," said Brown. "You're not pulling my leg about the odds?"

"Indeed I'm not," said Jones.

So Brown paid $1. **Was he well advised to do so?**

89 DEAD ON TIME

"Have you heard about young Speedmore and his bet?" asked Professor Probe at the Clownchester Club.

"No," I said. "I don't even know who young Speedmore is."

"He's one of our demonstrators," said Probe. "What he particularly likes demonstrating is his prowess on wheels. Last week he happened to mention that his new car was at Coventry, which is just 300 miles from here. And the upshot of a lot of backchat which followed was that Speedmore laid a bet to this effect: that he would cycle to Coventry; transfer himself to his car; drive to Queen's Kirby—which is 300 miles from Coventry and also 300 miles from Clownchester—and be back here within 24 hours."

"And he won the bet?"

"Just," said Probe. "He arrived dead on time. His average speed between Coventry and Queen's Kirby was exactly three times his average speed between Clownchester and Coventry; then—finding himself behind time—he put on another 15 miles per hour (on the average) over the third leg of his triangle."

What was Speedmore's average speed on his bicycle?

90 CUTTING FOR PARTNERS

Professor Probe and three of his cronies were cutting for partners in a bridge game. Each drew a card of a different suit.

"Just what one would expect," remarked Colonel M'Topee. "That's right, isn't it, Professor?"

"Well, no," said Probe. "You'd lose a lot of money if you laid odds on all four suits appearing."

"You'd lay odds against it?"

"Indeed I would," said Probe. "I'd give you six or seven to one. In fact, my dear Colonel, you could safely put even money on there being two cards of one suit and one each of two others. In the long run, you'd be a handsome winner."

"Lay me half a dollar on it next time," said the skeptical M'Topee.

"I'll do so each rubber," said Probe. They played five more rubbers, and the Professor won the odd half-dollar.

What, in fact, are the odds that, when four players cut for partners, two cards of one suit are exposed, and one each of two others?

91 SAVINGS CERTIFICATES

"I've just bought 2,000 Savings Certificates," said Sir Henry, whom I met coming out of his bank.

"Nice work," I said. "I didn't know they'd sell you so many."

"Ah!" said Sir Henry. "But they're not for me. Fifty-one of them are for my children. The other 1,949 I'm giving to my nephews and nieces—to celebrate the New Year."

"It's a prime number, isn't it?" I said. "1,949, I mean. So some of your relatives will be luckier than others."

"Yes," said Sir Henry. "The boys will be getting more than the girls. Each boy, in fact, will get five more certificates

than each girl. Also—if you want to make a puzzle out of it—
my nephews outnumber my nieces by three."

How many certificates does each nephew receive?

92 MARGARINE

"There you are," said Mrs. Probe to the Professor; she had
set out a number of samples on a plate. "If you can find three
which you are certain are butter I'll stand you dinner and the
flicks."

"I'm not much good at this sort of thing," said Probe.
"They all look alike to me. I might just as well choose three
at random. Are half of them butter and half margarine, by the
way?"

"No. The pats of margarine outnumber the pats of butter
by two."

"That so?" said Probe. He made a rapid calculation. "If
you asked me to choose three margarines, and I made my
selection at random, I should stand some small chance of my
free dinner. The odds would only be five to one against my
being right."

"Too easy," said Mrs. Probe. "You see if you can't pick out
three butters."

**If Probe makes a random selection of three samples,
what are the odds against their all being butter?**

93 HEADACHE FOR THE CITY SURVEYOR

The employees of the Clownchester Corporation put forward
demands recently for a shorter working week and better pay.

The City Surveyor received an employees' deputation.
"Kindly let me know," he said, "exactly what it is that you
want."

The spokesman of the deputation produced a memorandum.

"Sir," he said, "we are paid, you will recall, by the hour, and we work five days a week. We are putting forward two demands. (1) Our working day to be in future one hour shorter. (2) Our hourly rate to be raised to a point at which, in the course of the shorter working day that we are demanding, we can earn as many dollars as we should have to work hours in order to earn $49. At present, of course, a full day's work brings in one dollar less than that—and incidentally as many dollars as it is necessary to work hours in order to earn $40."

"I see," said the City Surveyor, mendaciously.

For how many hours a day are the employees proposing to work; and at what rate of payment per hour?

94 GRAND CHAIN

Alice, Betty, Clara, and Dolly are playing a simple game with a set of 28 dominoes. They have seven dominoes each to begin with, and, starting with Alice, take turns (in the order set out above) in the building of a chain.

It is a condition that one end of each domino played must be numerically identical with one of the open ends of the chain that is being built.

The total "pip values" of the first two dominoes played by each are: Alice's two dominoes: 23, Betty's two dominoes: 20, Clara's two dominoes: 18, Dolly's two dominoes: 16 .

On the third round, Alice plays the 6–2.

What, in sequence, are the first eight dominoes played?

95 TEDDY BEARS

"Our house is full of Teddy bears," said Boniface when I saw him at the club. "They can be got in three sizes, as perhaps you know. We've been buying 'em—Betty and I—for distribution to the hospitals.

"We spent three days getting the collection together. The first day, I concentrated on buying the smallest size, and Betty on buying the largest. The second day, we both bought middling-sized ones. And the third day, I was buying the big chaps and Betty the little ones.

"Very odd, our shopping was—from an accounting point of view. I bought one bear fewer on the second day than on the first, and one bear fewer again on the third day. With Betty, it was the other way round. She bought one more bear on the second day than on the first, and one bear more on the third day than on the second.

"As to price, there wasn't a great deal in it. Each of us paid 20 cents more for a middling-sized bear than for a small one, and 20 cents more for a big bear than for a middling-sized one. And—believe it or not—we each spent the same amount: $28.67."

"Then you must have paid the same prices for your bears," said someone; "also you bought equal numbers of each."

"Neither conclusion follows," said Boniface a trifle testily. "Betty's a better shopper than I am, and contrived to buy more bears.

"And, by the way, Caliban" (he added), "you'll need to know—if you want to make a problem of it—that each size of bear costs so many cents exactly."

How many more bears were bought by Betty than by Boniface?

96 RIPCHESTER

I was present this year at the athletic sports at Ripchester. The three houses—Walpole, Chatham, and Peel—compete one against another. There are three representatives of each house in each race. The first home scores one point; the second, two points; and so on; the house with the lowest aggregate wins that particular race.

The mile was a thrilling affair. My nephews, George and James, were running for Chatham. They are identical twins and rivalry between them is keen. Their elder brother, Tom, was Chatham's first string; he came in well ahead of the twins. But George and James were neck and neck all the way—hoping, I believe, to dead-heat—and George wasn't more than six inches in front of James.

This race produced a triple tie, all three houses clocking up the same number of points.

In what position did Tom finish?

97 A LOSING BATTLE

"I've come out second best in my battle with the men's union," said Sir Jonas Whalebone disgustedly.

"How so?" asked someone.

"It was a matter of getting some thousands of crates shifted," said Sir Jonas. "The exact number"—he consulted his notebook—"was 69,489; and the job took nine working days. I didn't think the chaps shifting them were putting all they had into it. The union leaders thought otherwise. Every day after the first I put six more men onto the job; and every day after the first each man—by arrangement—shifted five fewer crates than was the quota for the day before. The result was that, during the latter part of the period, the daily number of crates shifted actually began to go down."

"You should have tried payment by results," said someone.

Whalebone snorted. "Try telling that to my union."

What was the largest number of crates shifted on any one day?

98 THE THIRTY-SEVENTH INTERNATIONAL

"I'm boss of the Thirty-seventh International," said my friend Gallowglass.

"What on earth does that do?"

"Between ourselves, nothing much," said Gallowglass. "But it helps a lot of chaps to feel important. That's something these days, when we're all pushed around so much."

"How many more of you are there?"

"There are thirty-six Panjandrums," said Gallowglass. "In addition to me, the Grand Panjandrum. Each of them has a number—Number One, Number Two, and so on up to Number Thirty-six, and I have got them organized in twelve groups of three: Recruitment; Propaganda; Ideology; Background; Continuity; Liaison; Research; Experiment; Statistics; Teleology; Integration; Therapeutics. Pretty useful, what?"

"If you say so. But why are you telling me?"

"We've a vacancy," said Gallowglass. "Number 36 was run over by a traction engine. Five hundred a year and expenses."

"H'm," I said doubtfully. "I know nothing about Therapeutics."

"You wouldn't be Therapeutics. You'd be Experiment. Number 36 works with 11 and 27. All that's arranged on an intelligible plan which it's easy for me to memorize. Number One's colleagues are 10 and 26. Number Two's are 15 and 20; and so on. With a few minutes' thought you could work out the arrangement for yourself."

What are the numbers of Number Three's two colleagues?

99 THE INGENIOUS MR. DRAKE

For the individual Bowls Championship at our club the ingenious Mr. Drake (our secretary) divided the competitors into Sections A, B, and C. There were 30 entrants in all.

"Which means, I presume" (said someone) "that there will be 10 players in each section?"

"Not at all," said Mr. Drake. "I've tried to give everyone

a chance of winning a prize. There aren't ten players whom I can properly put in Section A."

"Then how many are there in Section A?"

Drake consulted his notebook. "Two fewer than in Section B."

"You are a maddening fellow. How many in Section B?"

Drake studied his calculations again. "Each competitor," he said, "will be playing one game against every other competitor in his section. That means that in Sections A and B, taken together, seven more games will be played than in Section C."

So how many competitors are there in each of the three sections?

100 THE LOCAL BOTWINNIKS

"The local Botwinniks will be on parade next week," said the secretary of the Gloomshire Chess Club.

"And what does that mean?" I asked.

"We're holding our annual Chess Congress. Quite a nice entry there is, too—only two fewer than last year."

"Are all the entrants playing one another?"

"Good heavens, no. Last year they were organized in a convenient number of sections, with the same number of players in each section. This year there will be one more section than was the case last year, but the number of players in each section has been reduced by one."

"How many games in all will be played?"

"I must work that out," said the secretary. "It's 126 fewer than was the case last year: that much I can tell you. Each player, by the way, plays two games against each of the other players in his section."

How many competitors are there?

SOLUTIONS

1 CINDERELLA

If a, b, and c are the three sisters' ages, Cinderella's being a, the data give:

$$(b+1)(c+1) + 2(b+2)(c+2) = 1382$$
$$\text{i.e., } 3bc + 5b + 5c = 1373$$

multiply by 3 and add 25

$$\text{and } 9bc + 15b + 15c + 25 = 4144$$
$$\text{i.e., } (3b+5)(3c+5) = 4144$$

Resolving 4144 into factors, three possible answers emerge, only one of which is reasonable—i.e., 17 and 23.

Therefore, the age of Cinderella is 15 years.

2 MIRANDA TURNS THE TABLES

Let M, E, L, D, S be the ages of Miranda, Eva, Lucinda, Dorothea, and Stella respectively. Then we have the equations:

$$M + E + L + D + S = 5M \quad (a)$$
$$M + (3M - S) + D + (3M - S) = 5M \quad (b)$$
$$E + (3M - S) = 3E \quad (c)$$
$$L + (3M - S) = 2S + 1 \quad (d)$$

$$(c) \text{ gives } 2E = 3M - S$$
$$(d) \text{ gives } L = 3S - 3M + 1$$
$$(b) \text{ gives } D = 2S - 2M$$

So, substituting in (a)

$$2M + 3M - S + 6S - 6M + 2 + 4S - 4M + 2S = 10M$$
$$\text{i.e., } 11S + 2 = 15M$$

This equation is satisfied by $S = 8$; $\quad M = 6$
$$S = 23; M = 17, \text{ etc.}$$

Since Miranda has just taken School Certificate, the required solution is that which makes her 17 years old.

Miranda is therefore 17.

3 FUN ON THE *STYGIAN*

1. Four solvers divide 5 points daily. There are six possible divisions of these five points:

        (1) 5–0–0–0   (2) 4–1–0–0   (3) 3–2–0–0
        (4) 3–1–1–0   (5) 2–2–1–0   (6) 2–1–1–1

2. From the above can be calculated how much changes hands in respect of each division of points:

| | $ Won or Lost by Solvers | | | | Total $ |
Division	1	2	3	4	Changing Hands
(1)	15	−5	−5	−5	15
(2)	11	−1	−5	−5	11
(3)	7	3	−5	−5	10
(4)	7	−1	−1	−5	7
(5)	3	3	−1	−5	6
(6)	3	−1	−1	−1	3

3. We have now to construct a "profit and loss" account for the five solvers. We know that A has won $20 and that D and E have each lost $12. Now $12 can only be made up by taking $5 twice and $1 twice; hence Division (6) must occur twice; this accounts for $6 out of $41.

4. The balance of $35 can only be made up of one $15 and two $10's—i.e., No. (3) occurs twice and No. (1) once. A's $20 win must be made up of two $7's and two $3's.

5. The table can now be quickly completed:

Profit and Loss in $

Setter	A	B	C	D	E
A	—	15	−5	−5	−5
B	3	—	−1	−1	−1
C	3	−1	—	−1	−1
D	7	3	−5	—	−5
E	7	3	−5	−5	—
Totals	20	20	−16	−12	−12

Biffins wins $20; Chump loses $16.

4 VILLAGE BAZAAR

1. The number of children cannot have been six or fewer. For they have $70 to spend altogether; no two may spend the same amount; and the most any child can spend is $13.

2. Also there cannot be 8 children. For in this case the most favorable distribution of their money (5, 6, 7, 8, 9, 10, 12 and 13 dollars) involves expenditure at more than three stalls.

3. *A fortiori*, there cannot be more than 8.

4. Hence there were 7 children; and it can quickly be ascertained (by trial) that the only expenditure conforming to the data is:

$ 7	(2, 2, 3)	$11	(3, 3, 5)
$ 8	(2, 3, 3)	$12	(2, 5, 5)
$ 9	(2, 2, 5)	$13	(3, 5, 5)
$10	(2, 3, 5)		

Seven children went to the Bazaar, and patronized the 2-, 3-, and 5-dollar stalls.

5 THE JONES BOYS

Let the ages of the boys be $x+1$, x, $x-1$ years.
Then they receive collectively:

$$(x+1)x + (x+1)(x-1) + x(x-1) \text{ dollars,}$$
i.e., $(3x^2 - 1)$ dollars.

Similarly, if their ages on the occasion of my previous visit were $y+1$, y, $y-1$ years, they then received $(3y^2 - 1)$ dollars.

$$\text{Hence } 3x^2 - 3y^2 = 120$$
$$\text{i.e., } (x+y)(x-y) = 40$$

This gives x 11, y 9; or x 7, y 3.

The wording of the problem clearly renders the second solution inadmissible.

Hence two years elapsed between my two visits.

6 APRIL'S BROOD

April has thirteen children—five boys and eight girls.
Each boy has 40 dollars to spend, and they invest it at the
various stalls as follows:

Prices of Articles Sold in Dollars

Boys	1	2	3	4	5	6	7	8	9	10	11	12
A	x	x	x	x	x	x	x					x
B	x	x	x	x	x	x		x			x	
C	x	x	x	x	x	x			x	x		
D	x	x	x	x	x		x	x		x		
E	x	x	x	x		x	x	x	x			

There is no other solution consistent with the data.

The problem can most easily be solved by trial of the relevant
factors of 200.

7 INTELLIGENCE TESTS

Let there be n tests and p girls participating.

Then the number of cents paid out in respect of each test is

$$17 - (p-1) = 18 - p.$$

Now altogether 360 cents are paid out, so $n(18-p) = 360$ (1)

Let x be the number of tests won by Joy.

Then $(n-x) - 17x = 30$, or $n - 18x = 30$ . . . (2)

Hence n is a multiple of 6, say $6r$.

Then $r = (3x+5)(18-p) = 60.$
       or $3x(18-p) = 5(p-6)$ . . . (3)

As all the letters represent integers, we deduce from (3) that p
is greater than 6 and less than 18, and that $(p-6)$, and therefore
p also, is a multiple of 3.

The only possible values of p are 9, 12, 15; and of these only 15 gives an integral value of x.

Thus $p = 15$, $x = 5$, $r = 20$, and $n = 120$.

Hence there were 15 participants and 120 tests, the numbers won being 1, 2, 3 . . . 15 respectively.

Joy won 5 tests.

8 BRIDGE

The results of the five rubbers are as follows:

1st rubber: Colonel and Doctor beat Padre and Admiral.

2nd rubber: Doctor and Padre beat Colonel and Admiral.

3rd rubber: Colonel and Admiral beat Doctor and Padre.

4th rubber: Colonel and Doctor beat Padre and Admiral.

5th rubber: Colonel and Padre beat Doctor and Admiral.

An analysis of *possible* winning scores gives the following (in dollars).

79	61	53	47	43	35
33	29	25	21	17	15
11	7	3	1		

But two scores must be selected which show a difference that is a multiple of $15.

These are:

$$61 : 1$$
$$47 : 17$$
$$33 : 3$$

But only in the first case are the conditions regarding losses fulfilled. 61 (i.e., $18 + 23 + 13 + 16 - 9$) $- 1$ (i.e., $18 + 9 + 13 - 23 - 16$) $= 15[33$ (i.e., $23 - 18 - 9 - 13 - 16$) $- 29$ (i.e., $9 + 16 - 18 - 23 - 13$)].

Winnings and losses are therefore as set out above.

9 DODECAHEDRA

The number of distinguishable dodecahedra is 96, subdivided thus, in respect of color distribution:

Faces		
12–0	. . . .	2
11–1	. . . .	2
10–2	. . . .	6
9–3	. . . .	10
8–4	. . . .	24
7–5	. . . .	28
6–6	. . . .	24
Total	. . .	96

10 THE SIMIAN LEAGUE

Let the number of clubs last year $= a$, and number added this year $= b$.

Matches played last year $= a(a-1)$, this year $(a+b)(a+b-1)$.

Then $(a+b)(a+b-1) - a(a-1) = 16b$
i.e., $b^2 + 2ab - b = 16b$
$\therefore b + 2a - 17 = 0$ (or $b = 0$)
$\therefore b = 17 - 2a$

Also $(a+b+1)(a+b) - (a+b)(a+b-1) - a(a-1) = 4$
i.e., $-a^2 + 3a + 2b = 4$

Substitute $(17 - 2a)$ for b.
$$-a^2 - a + 30 = 0$$
$$\therefore a = 5 \text{ or } -6.$$

Positive root only is permissible, and $b = 7$.

Number of clubs last year = 5; this year = 12.

11 THE CROSS-COUNTRY FINALS

If there are a runners in Class A, and b runners in Class B, in a given year, the total number of points lost is ab. When this has been proved, the next step is to find all possible A scores and all possible B scores, selecting those that occur two or more times.

It is possible that the value of ab might have been the same in two of the years, but this is found not to be the case, and we need only consider the scores that occur three times.

In one case only do the corresponding values of ab add to 17. The values of a and b in this case are:

$$(9, 1) \ (2, 4) \ \text{and} \ (0, 9).$$

Sprintwell was once in Class A and twice in Class B, with a plus score of 2 each time.

12 GOOD EGGS

Humpty Dumpty's mark in arithmetic was 10.

There were altogether seven Good Eggs who qualified, because the number of marks necessary to qualify was $4 \times 3 \times$ the number of subjects, and at the same time twice as many as the number of Good Eggs besides Humpty Dumpty, who qualified, multiplied by the number of subjects.

There must have been at least five subjects, because the total number of marks necessary to qualify was four times the maximum obtainable in one subject and no marks were repeated in any one score.

In the case of five subjects, there are exactly seven ways to score:

$$60 = 15 + 14 + 13 + 12 + 6$$
$$60 = 15 + 14 + 13 + 11 + 7$$
$$60 = 15 + 14 + 13 + 10 + 8$$
$$60 = 15 + 14 + 12 + 11 + 8$$
$$60 = 15 + 14 + 12 + 10 + 9$$

$$60 = 15+13+12+11+9$$
$$60 = 14+13+12+11+10$$

In the case of more than five subjects, there are many more than seven ways to score. **Humpty Dumpty's mark in arithmetic, therefore, was 10.**

13 GRINDGEAR'S REGISTRATION NUMBER

Let the sum of the primes referred to be m.

Let $a = m-1$, $b = m+1$, $c = 100-2m$.

We now have the equation:

$$n\left[\frac{1}{m-1}+\frac{1}{m+1}+\frac{1}{100-2m}\right] = E,$$

where E is the product of the four primes whose sum is m.

The equation reduces to $n(200m-3m^2-1) = E(m-1)(m+1)(100-2m)$.

Now any factor of $(200m-3m^2-1)$ and $(m-1)$ is also a factor of $(200m-3m^2-1)+(3m-1)$—i.e., of $196m$.

Similarly any factor of $(200m-3m^2-1)$ and $(m+1)$ is also a factor of $204m$, and any factor of $(200m-3m^2-1)$ and $(100-2m)$ is also a factor of (m^2-1). Now m is at least 17 and at most 47.

Also as $196 = 4 \times 49$ and $204 = 4 \times 51$, the only factors of $(200m-3m^2-1)$ which could occur in $(m-1)(m+1)(100-2m)$ are 3, 4, 7, and 17. The other factors must be sought in E, in which the largest possible factor is 37.

Thus in testing the values of this expression for values of m from 17 to 47 we reject at once all which have prime factors greater than 37. The prime factors of those which remain will contain some or all of the factors of the corresponding E.

It will be found that the only possible value of m is 29, leading to $(200m-3m^2-1) = 3276 = 4 \times 7 \times 9 \times 13$.

Of these factors 4, 3, 7 are divisors of $(m-1)(m+1)(100-2m)$.

Two of the factors of E must be 3 and 13; the others can only be 2 and 11 since the sum of the four factors is 29. **Hence Grindgear's number is 9240.**

14 THE SOCCER CHAMPIONSHIP

Anderson, Barnes, and Dickens	.	0	*vs.*	Chaplin, Egerton, and Fowler	.	3
Anderson, Barnes, and Fowler	.	4	*vs.*	Chaplin, Dickens, and Egerton	.	0
Anderson, Chaplin, and Dickens	.	2	*vs.*	Barnes, Egerton, and Fowler	.	0
Anderson, Chaplin, and Fowler	.	1	*vs.*	Barnes, Dickens, and Egerton	.	2
Anderson, Dickens, and Egerton	.	3	*vs.*	Barnes, Chaplin, and Fowler	.	1
Anderson, Egerton, and Fowler	.	0	*vs.*	Barnes, Chaplin, and Dickens	.	1

Solution: With six players there must be 10 matches, and since 23 goals were scored with only two scores the same, it is seen that the scores were: 0–0, 0–0, 1–0, 2–0, 1–1, 3–0, 2–1, 4–0, 3–1, and 2–2.

Since the winner had more goals scored against him than for him, and since the loser had more goals for than against, it is safe to assume that they were all very close together, that the winner had 11 goals for and 12 against, and that the loser had 12 for and 11 against. The three who were bracketed must have scored 11 for and had 12 against, and the remaining player scored 13 for and 10 against.

The following table is now constructed:

	Possible Scores										Goals		Matches			Pts.
	0–0	0–0	1–0	2–0	1–1	3–0	2–1	4–0	3–1	2–2	For	Agst.	W.	D.	L.	
D	0–0	0–0	1–0	2–0	1–1	0–3	2–1	0–4	3–1	2–2	11	12	4	4	2	12
A	0–0	0–0	0–1	2–0	1–1	0–3	1–2	4–0	3–1	2–2	13	10	3	4	3	10
C	0–0	0–0	1–0	2–0	1–1	3–0	1–2	0–4	1–3	2–2	11	12	3	4	3	10
B	0–0	0–0	1–0	0–2	1–1	0–3	2–1	4–0	1–3	2–2	11	12	3	4	3	10
E	0–0	0–0	0–1	0–2	1–1	3–0	2–1	0–4	3–1	2–2	11	12	3	4	3	10
F	0–0	0–0	0–1	0–2	1–1	3–0	1–2	4–0	1–3	2–2	12	11	2	4	4	8

Having completed the table satisfactorily, it is seen that—

in the 1–0 match D, C, and B beat the others;
in the 2–0 match D, A, and C beat the others;
in the 3–0 match C, E, and F beat the others;
in the 2–1 match D, B, and E beat the others;
in the 4–0 match A, B, and F beat the others;
and in the 3–1 match D, A, and E beat the others.

15 TRIANGLE GOLF

We can classify all possible results of the 5 holes' play.

Now, since Slicer loses each day, and each day after the first loses more, while Fluff each day wins only two holes, the following must be the distribution of wins which successively apply:

15,   14,   13,   9,   8,   6,   7,   5

—these being the only distributions which allow Fluff to win two holes.

	Holes Won by:			Won or Lost in $		
No.	a	b	c	a	b	c
1	5	− 0	0	+30	−15	−15
2	4	1	0	+19	− 8	−11
3	31	1	0	+13	− 5	− 8
4	22	1	0	+11	− 4	− 7
5	3	2	0	+ 9	—	− 9
6	21	2	0	+ 5	+ 2	− 7
7	3	11	0	+10	− 2	− 8
8	21	11	0	+ 6	—	− 6
9	111	11	0	+ 4	+ 1	− 5
10	3	1	1	+10	− 5	− 5
11	21	1	1	+ 6	− 3	− 3
12	111	1	1	+ 4	− 2	− 2
13	2	2	1	+ 2	+ 2	− 4
14	2	11	1	+ 3	—	− 3
15	11	11	1	+ 1	+ 1	− 2

Now we can consider what happens at each hole:

Day	Distribution			Winner of Holes				
				1	2	3	4	5
1	11	11	1	d	f	d	S	f
2	2	11	1	S	f	d	d	f
3	2	2	1	f	f	D	d	S
4	111	11	0	d	f	d	f	d
5	21	11	0	d	f	d	d	f
6	21	2	0	d	f	f	d	d
7	3	11	0	f	d	d	d	f
8	3	2	0	f	f	d	d	d

The data are given in the second table in capital letters. It is at once obvious also that Divot won the 2nd, 3rd, and 4th holes on the 7th day, Fluff taking the 1st and 5th, and that Divot won the 1st, 3rd, and 5th holes on the 4th day. From this it follows that the hole won seven times by Fluff was the 2nd. The remaining wins can now be readily deduced.

Hence Fluff won the following holes:

1st day . . .	**2nd and 5th**
2nd „ . . .	**2nd and 5th**
3rd „ . . .	**1st and 2nd**
4th „ . . .	**2nd and 4th**
5th „ . . .	**2nd and 5th**
6th „ . . .	**2nd and 3rd**
7th „ . . .	**1st and 5th**
8th „ . . .	**1st and 2nd**

16 TALL SCORING

Ten goals were scored in each match. No two matches produced the same result. So the six results were:

(a) 10–0; (b) 9–1; (c) 8–2; (d) 7–3; (e) 6–4; (f) 5–5.

We have two known results, as well as the table of goals "for" and "against." F drew with B 5–5, and defeated Q 8–2.

So F lost 4–6 to Snug. Snug had a second win, which must have been 7–3. If this was against Q, Q scored eight goals in their third match, which is impossible. Hence Snug won 7–3 against Bottom, and lost 0–10 to Quince. This leaves the 9–1 game unaccounted for.

Bottom beat Quince by nine goals to one.

17 BOWLS

If it's an even-money chance against a player winning the toss m times (at least), he must be playing $(2m-1)$ matches.

For he is equally likely to win the toss every time (i.e., $2m-1$ times) and not at all; he is equally likely to win it $(2m-2)$ times and once . . ., and so on to m and $(m-1)$ times. Take the case of seven matches. The toss can be won in 2^7, i.e., 128 ways. This total can be analyzed as follows:

D wins	7	6	5	4	3	2	1	0
No. of ways	1	7	21	35	35	21	7	1

Drake and Raleigh are playing seven matches.

18 MRS. COLDCREAM OBJECTED

We must first determine how many girls there are.

If there are m girls, each of whom receives $\$x$, we get the equation:

$$mx + (19 - m)(x + 30) = 1,000$$

This resolves itself into:

$$19x - 430 = 30m$$

Since m is less than 19, there is only one integral solution of this equation, i.e., $m = 11$, $x = 40$.

The committee had proposed to award $\$70$ to each of eight boys, and $\$40$ to each of 11 girls.

So, as a result of Councillor Coldcream's protest, **each boy received $\$59$, and each girl received $\$48$.**

19 FOOZLEDOWN

Let there be m players in all.

Then the number of possible fours is mC4, and each four provided three different matches.

So mC4 = 15; whence m = 6.

So there were four players in addition to Deadpan and Backspin.

20 CUBES

(1) We can have four black lines forming one continuous line (a square) all round the cube. Then (A) the other two lines can be parallel with one another; or (B) the other two lines can be at right angles to one another.

(2) We can have three black lines forming a continuous line (three sides of a square). Then (C) there can be a second system of three black lines, the middle line of which is on the opposite face to that carrying the middle line of the first system, and at right angles to it. Or (D) there can also be a system of two lines (two sides of a square) and one line which is isolated. Or (E) the three other lines can all be isolated.

(3) There can be a system of three pairs of lines, each pair forming two sides of a square. Two such systems are possible (F and G); one is the "mirror reflection" of the other.

(4) (H) All six lines can be isolated. A–H cover all possibilities.

So there can be eight distinguishable cubes in all.

21 MORE CUBES

The simplest solution of this problem is based on the previous solution. There are eight distinguishable cubes having a central line parallel to two edges. If, in each of these, we imagine every central line to be rotated 45 degrees to the right, as viewed from a point within the cube, we must still have eight distinguishable cubes which will satisfy our new conditions. And there can be no others.

So the answer, again, is eight.

22 TABLE TENNIS

This is not a difficult calculation.

The "state of the score" can be best understood from the table:

	S	L	W	C
S	–	1	1	1
L	0	–	1	1
W	0	0	–	1
C	0	0	0	–

Each player has three more games. So to win the trophy outright W must win all three games (for S has won three already); S must lose to L and C, as well as to W; and L, having lost to W and beaten S, must lose to C. The score will then be: W 4; S 3; L 3; C 2; and this is the only combination of results which will enable W to win his bet.

These results cover all six games, and the chance of each is $\frac{1}{2}$. So the chance of all six results favoring W is 1–64.

The odds against Wallaby are 63 to 1, and he should not accept Faredooze's offer.

23 KINDHARTZ

The Kindhartz formula can only be determined by trial. A clue is given by the fact that the pension awarded at 80 is double that awarded at 60. This suggests that the pensioner's age is deducted from 100, and that some agreed number is then divided by the difference.

It should not be difficult to deduce that this number is 25,200. If a pensioner's age on retirement is m years, his pension in dollars will be found to be $\frac{25,200}{(100-m)}$. So the employee in question has retired at 64; had he waited for another year, **his pension would have been $720.**

24 DOWNING STREET

Call the two speakers A and B. If the same 20 questions are put to both of them, the chances are that they will both give truthful answers to 12 of them; contradictory answers to 7 (3 plus 4); and untruthful answers to one. So, for every 13 times they agree on an answer, it is likely to be incorrect on one occasion.

Hence the odds are 12 to 1 on that it actually was the Minister.

25 CRANKSHAFT

(1) Gearbox covers 39 miles while Crankshaft covers 13. So Gearbox is traveling three times as fast as Crankshaft.

(2) Hence, when Gearbox left the Stork, Crankshaft was 10 miles away, i.e., he had already covered 16 miles. Let his speed be m miles per hour.

Then $\dfrac{16}{m} + \dfrac{52}{3m} = \dfrac{10}{3}$.

(3) So Crankshaft's speed is 10 m.p.h., and

He was back at the Three Tuns at 5:12 P.M.

26 MARBLES

The first operation is simple enough. Five marbles must be transferred: for the first four drawn may all be of the same color.

Now consider the second operation. There are four possibilities:

MARBLES LEFT IN BAG 1			MARBLES NOW IN BAG 2		
a	b	c	a	b	c
0	3	4	8	5	4
1	2	4	7	6	4
1	3	3	7	5	5
2	2	3	6	6	5

In each case it may be necessary to retransfer as many as 12 marbles. For in the first case the first nine marbles to be retransferred may all be *b* and *c*. And similarly with the other cases.

So there will be five marbles left in Bag No. 2.

27 MUCH SPENDING

The area of the two triangles is five-twelfths the area of the estate. So the shortest side of each triangle is 5/12 of the side of the square; hence the longest side of each is 13/12 of the side of the square, i.e., of 1,320 yards. The central avenue is, of course, the same length.

So the central avenue is 1,430 yards long.

28 ALGEBRAICA

1X is in our notation 17 (13 plus 4).

17^2 is in our notation 289.

289 = 169 plus 120.

169 is 100 in the notation of Algebraica, 120 is 73 (for in our notation 120 is 9×13 plus 3, and in Algebraica's notation 9 is 7).

So the square of 1X is 173.

29 CLOWNE WAS UNLUCKY

It's a simple calculation, though not quite so simple as Clowne supposes. The chance of his failing to win is 4–5 each day. So the chance that he fails to win five days in succession is:

$$\frac{4}{5} \times \frac{4}{5} \times \frac{4}{5} \times \frac{4}{5} \times \frac{4}{5} = \frac{1024}{3125}$$

or odds of 2101 to 1024 against.

30 LITTLE MATING

If there are m players in a section, $\dfrac{m(m-1)}{2}$ games are played.

Let there be m players in Section B.

Then $\dfrac{m(m-1)}{2} - \dfrac{(24-m)(23-m)}{2} = 69.$

Whence $m = 15$.

So there were 9 players in Section A.

Mr. Gambit drew 5 games.

31 TETRAHEDRAL

(1) On each face, clearly, ABC must be painted in one color, and D in a second.

(2) Call the four colors M, N, P, Q. Then each must be used for three triangles on one face. Two different arrangements, however, are possible. E.g., if the tetrahedron stands on base M, the other three faces in clockwise order can be NPQ or NQP.

(3) With each of these arrangements, nine variations of the second colors of each face are possible. For we can have:

MN	NP	PQ	QM
MN	NQ	PM	QP
MN	NM	PQ	QP

and other similar variations for MP and MQ.

So my tetrahedron can be painted in 18 different ways.

32 PEDAL AND HOOFIT

Hoofit's cycling time is 2/5ths of six hours, i.e., 2 2/5 hours.
Pedal's cycling time is $\frac{1}{4}$ of six hours, i.e., $1\frac{1}{2}$ hours.

It follows that Hoofit's speed on his bicycle is 5/8ths of Pedal's speed on a bicycle; i.e., Pedal, when the two cyclists pass one another, has 8/13ths of the distance still to cover.

So they pass one another 10 miles from the Anchor.

33 MRS. INKPEN

A puzzle within a puzzle. Neither is very difficult.

If Patriarch had m grandsons, each of whom received $\$n$, we. have:

$$mn + (31 - m)(n + 7) = 470$$
$$\text{i.e., } 31n - 7m = 253$$

There is only one solution of this equation in positive integers with m less than 31, $m = 17$; $n = 12$. So 17 grandsons each received $\$12$ and 14 granddaughters each received $\$19$.

Hence $\$74$ is the amount received by three grandsons and two granddaughters.

So Mrs. Inkpen had two daughters.

34 THE TOOTLES

The Tootles forgot that two throws of a dice give varying chances of different totals. These varying chances are:

TOTAL	2	3	4	5	6	7	8	9	10	11	12
CHANCES	1	2	3	4	5	6	5	4	3	2	1

Taking this factor into account, we can readily ascertain the theoretical distribution of the chairs counted out in 36 throws. For example, 7 (the most likely throw), counts out A and C; 6 counts out F and A. The sum of all chances gives:

A	B	C	D	E	F
15	10	15	11	11	10

The Tootles had therefore made the worst possible selection. **They should have chosen seats A and C.**

35 HELPUSELPH

The two sides of the rectangle total 23 miles. Hence, if m miles be one side of the rectangle,

$$(m + 4)(23 - m + 5) = 2m(23 - m)$$

So m is either 14 or 8.

The Governor had had in mind a rectangle 15 miles by 8 miles (which is half the area of a rectangle 20 miles by 12 miles). The applicant selected a rectangle 14 miles by 9 miles (which is half the area of a rectangle 18 miles by 14 miles).

So the area in question was 126 square miles.

36 DROPPED CARD

The card which is dropped is equally likely to be any one of 13. Four of these are known to be Hearts. Each of the remaining 9 may be any one of 48 cards, of which 9 are Hearts.

So the chance that the card dropped is a Heart is:

$$\frac{4}{13} + \frac{9}{13} \times \frac{9}{48}$$

$$\text{i.e.,} \ \frac{273}{13 \times 48} \ \text{or} \ \frac{7}{16}$$

The odds are 9 to 7 against the card dropped being a Heart.

37 HIGHTONE

Emerson lost 3–8 to Ruskin, so they scored 15 goals in the two matches they won. Hence they won these matches 6–5 and 9–2. This, it will be found, gives us two possibilities consistent with all the data:

	I						II		
	R	E	B	C		R	E	B	C
R	–	8	4	1	R	–	8	1	4
E	3	–	6	9	E	3	–	9	6
B	7	5	–	11	B	10	2	–	11
C	10	2	0	–	C	7	5	0	–

The results of four matches are uncertain; but it is evident from the above that **Bacon beat Carlyle by 11 goals to 0.**

38 POLYCHROME

When G meets S he has done 6/11ths of the circuit, so his speed is 6/5ths of S's initial speed.

The two runners take the same time to complete the course. Let S's speed for the first two miles be m miles per hour.

Then $\dfrac{2}{m} + \dfrac{2}{m+4} = \dfrac{4 \times 5}{6m}$

Whence m is 8 miles per hour.

So each athlete takes 25 minutes.

39 CHESS

Suppose that m men played for Doomshire, and n men played for Gloomshire. Then we have:

$$\frac{3n}{5} + \frac{100-m}{3} + \frac{2m-2n}{3} = 51$$

i.e., $5m - n = 265$ and
hence $n = 5\,(m - 53)$

Now n must be more than 50; m must be greater than n; and $(100 - m)$ must be divisible by 3.

Having regard to these considerations, there is only one solution of the above equation: $n = 55$, $m = 64$.

So 64 men played for Doomshire.

40 SNATCH

At first blush, the data seem inadequate. But let's see:

(1) Since the red cards total 26, the difference in the numbers of red cards held by the two sides must be an even number. So Hobo holds an even number of black cards and, therefore, an odd number of red cards.

(2) Hobo holds the two red Aces. So he has either 3 or 5 red cards. If he has 7 or more red cards, Puffin has 14 or more black cards, which is impossible.

(3) If Hobo has 3 red cards, we can complete all four holdings as follows:

	Ho	Ha	P	S
RED	3	5	7	11
BLACK	10	8	6	2

(4) If Hobo has 5 red cards, he has 8 black cards. So P and S have 17 red cards. But P has 10 black cards, and, therefore, 3 red ones; and S would have 14 red cards, which is impossible. Hence there is only one solution.

Snatch holds two black cards.

41 PRIVATE ENTERPRISE

The starting point should be E's receipts in the fifth month. If in that month he sells m bicycles at $\$n$ each, he gets $\$nm$. Similarly, in the first month he gets $\$(m-16)(n+4)$; in the second month, $\$(m-12)(n+3)$; and so on to the ninth month, when his receipts are $\$(m+16)(n-4)$. Adding, we find

$$9mn - 240 = 3{,}153$$

i.e., mn is 377, of which the only factors are 29 and 13; so in that month E sold 29 bicycles at $13 each. The remaining calculations are very simple; it will be found that E's profits in successive months were: $130; 153; 168; 175; 174; 165; 148; 123; 90.

So he made his largest profit ($175) in the fourth month.

42 DIVIDING THE PACK

Suppose there are m red cards in Portion A.

Then there are $2m$ black cards in Portion A; $(26-m)$ red cards in Portion B; and $(26-2m)$ black cards in Portion B.

Hence $\dfrac{26-2m}{51-3m} = \dfrac{1}{3}$

When $m = 9$.

So there were originally 27 cards in Portion A, and 25 cards in Portion B.

43 TROGLODYTES

The three percentages add up to 67 per cent. But these include 137 members counted three times over. Hence (1,561 − 274) Troglodytes, i.e., 1,287, represent 33 per cent of the total.

And there are 3,900 in all.

44 SPEEDWELL

For those who don't care about algebra, this can readily be solved by trial. The formal solution begins with the equation

$$\frac{m+6}{27-m} + \frac{21-m}{m+6} = \frac{17}{10}$$

where m is the distance in miles between Cloudburst and the Red Lion.

This produces the equation

$$(m - 12)\,(37m - 273) = 0$$

So m is either 12 miles or $7\frac{14}{37}$ miles.

But in the latter case the Porpoise would be $13\frac{14}{37}$ miles only from Cloudburst, and so just not halfway.

Hence the Red Lion is 12 miles from Cloudburst.

45 SEVEN DIGITS

This is very simple if you know one thing: that the last digit of the fifth power of a number is the same as the last digit of the number itself. So the number first thought of ends in 7.

Now, the fifth power of 10 is 100,000.

The fifth power of 20 is 3,200,000.

The fifth power of 30 is 24,300,000.

Clearly then the fifth power of 27 will have eight digits; the seven-digit number must be the fifth power of 17 (which actually is 1,419,857).

Dingo's original number was 17.

46 GLOOMSHIRE

(1) Each team scored three points. Gruntle (3–3) defeated Jitters (5–5). So no team drew three games; each therefore won one, drew one, and lost one.

(2) Jitters lost 2–3 and drew 2–2. So on the third Saturday they played Much Moaning, winning 1–0.

(3) Hence Gruntle played Deadalive on the third Saturday. They either drew this match 0–0, or lost 0–1. But if they drew 0–0, Deadalive would have drawn two matches. Hence Gruntle lost 0–1.

So on the third Saturday, Much Moaning lost 0–1 to Jitters, and Gruntle lost 0–1 to Deadalive.

47 NO LUCK AT ALL

Let the first rubber be m 100's, and the original stakes n dollars per 100.

Then $mn + 3n(m-1) + 9n(m-2) + 27n(m-3) = 596$.

This equation, simplified, becomes

$$n(20m - 51) = 298$$

But $298 = 2 \times 149$, and 149 is a prime number, so n is either 1 or 2.

If n is 1, there is no solution. If n is 2, m is 10.

So the original stakes were $2 per 100, and **the first rubber was one of 1,000 points.**

48 FLOATING VOTE

Let the speed of the current be m miles per hour. Then Victor's speed downstream is $11m$ m.p.h., and his speed upstream is $9m$ m.p.h.

Let the distance covered by Victor before turning be d miles. Then $1/2m = d/11m + d/9m$.

Whence d is 99/40 miles.

Victor turned around after covering 2 miles 836 yards.

49 PUTWELL WAS NETTLED

This is not difficult. It will be found by trial that, while Foozleham might have won $1, $2, $3, $4, or $5 in a variety of ways, he can only end "all square" (when winning four holes) in one way. He wins the 12th and 13th ($3); loses the 14th, 15th, and 16th ($6); and wins the 17th and 18th ($3).

So at the end of 17 holes Foozleham was $2 down to Putwell.

50 WHOPPIT

Let t be Whoppit's average annual total over a period of x years.

Then $(t+300)\ (x-3) = (t-300)\ (x+3)$. Whence $t = 100x$.
So $tx - (t+300)\ (x-3) = x(100x) - 100(x+3)\ (x-3) = 900$.

I.e., **Whoppit's aggregate exceeded that of Bludgeon or Thrust by 900 runs.**

The above (algebraic) solution is less formidable than it appears—and you may well have found the answer by intelligent guesswork.

51 RABBITT

The car, when Rabbitt met it, would have reached the station in another six minutes. So Rabbitt had been walking for 30 minutes. Hence, had the car met Rabbitt at the station, he would have arrived $(30-6)$ minutes earlier at the point where he actually met the car.

So he would have arrived home at 5:36.

52 LARRY

This should not have taken you long to solve.

Larry's speed, when he is rowing with the current, is double his speed when rowing against it. Hence, if the speed of the current is m miles per hour:

$2(5-m) = (5+m)$
Whence m is 5/3.
The speed of the current is $1\frac{2}{5}$ m.p.h.

53 MRS. ANTROBUS

Let Mrs. A's age be m years, and her mother's age n years.
$n^2 - m^2 = (n+m)(n-m) = 2,720.$

Factorizing 2,720, we have the following possibilities:

	$(n+m)$	$(n-m)$	n	m
(1)	68	40	54	14
(2)	80	34	57	23
(3)	136	20	78	58

(1) is clearly impossible; (3) we may regard as invalidated by "Darling's" conversation.
So Mrs. Antrobus is 23.

54 DEEPDENE

(1) There were no drawn games; no two games produced identical scores; 17 goals were scored in all. So the results of the six games played can only have been: (a) 1–0; (b) 2–0; (c) 2–1; (d) 3–0; (e) 3–1; (f) 4–0.

(2) Catullus wins two games and loses one. Their matches therefore were (a), (c), (f).

(3) Who defeated Catullus 4–0? It will be found (by experiment) that the winners can only have been Horace, who lost matches (d) and (e). Now we can deduce the completed table:

	V	C	H	O
Vergil	–	0–1	3–0	2–0
Catullus	1–0	–	0–4	2–1
Horace	0–3	4–0	–	1–3
Ovid	0–2	1–2	3–1	–

So Vergil defeated Ovid by two goals to nil.

55 PEWTER

This looks difficult, but isn't.
Suppose that there are n nephews, and p pots.
Then, in respect of each pot, the amount distributed by Uncle James is

$$\$5(n-1) - 10; \text{ i.e., } \$(5n-15).$$

So $p(5n-15) = 510$, or $p(n-3) = 102$.

This equation has several solutions:

n:	4	5	6	9
p:	102	51	34	17

But the first is excluded by there not being 100 pots; the third and fourth by the fact that two nephews get 16 pots each and the others at least one. Hence there are five nephews and 51 pewter pots.
So Snoggins receives $175 less $160, i.e., $15.

56 SCHOLARSHIP

The marks awarded total 75. A gets 12, and B 14; so C gets 15; D, 16; G, 18. Now we can produce a skeleton table:

	GREEK	LATIN	ENGLISH	FRENCH	MATH.	TOTAL
A	1					12
B		1	3	3		14
C	3			1	2	15
D						16
G			2		1	18

A series of simple deductions completes the table. D gets 1 mark for English. G's unstated marks are all 5's. And so on. It will be found that Drudge's marks are: **Greek 4; Latin 3; English 1; French 4; Mathematics 4.**

57 FIVE RUBBERS

Let Tenace's third rubber be one of $m(100)$ points at n dollars per 100.

Then he successively loses (in dollars): $(m+6)$ $(n-2)$; $(m+3)$ $(n-1)$; mn; $(m-3)$ $(n+1)$; $(m-6)$ $(n+2)$.

Adding, we get the equation:

$$5mn = 1,445.$$

i.e., both m and n are 17.

So Tenace's successive rubbers cost him (in dollars) 345; 320; 289; 252; 209.

I lost the first two of these sums, and won the last three.

So I won $85.

58 ELGIN'S MARBLES

The odds of 11 to 5 means 5 chances in 16. Hence, if there are s marbles in one bag, of which m are red, and t in the other, of which n are red:

$$\frac{m}{s} \times \frac{n}{t} = \frac{5}{16}$$

So s and t, which total 18, are factors of some number which is a multiple of 16.

There are two possibilities: (1) s is 10 and t is 8; (2) s is 16 and t is 2. In the first case, we have five red marbles in each bag:

$$\frac{5}{10} \times \frac{5}{8} = \frac{25}{80} = \frac{5}{16}$$

In the second case there are 10 red marbles in the first bag, and one in the other:

$$\frac{10}{16} \times \frac{1}{2} = \frac{10}{32} = \frac{5}{16}$$

Hence, in case (1), there are five blue marbles in one bag and three in the other (15 chances in 80 of two blues). In case (2) there are six blue marbles in one bag and one in the other (6 chances in 32 of two blues). So in either case **the odds against two blue marbles are 13 to 3.**

59 MILD GAMBLE

In both cases the odds offered slightly favor Sharpwits, but there is very little in it.

(1) Three cards can be drawn from a pack of 16 in 560 different ways.

(2) Three cards of the same suit can be drawn in 16 ways. For there are four possible selections of three Spades, and similarly with each of the other suits. So the punter's chance of success is 16/560, or 1/35.

(3) Three cards of different suits can be drawn in 256 ways. There are four possible suit combinations (S H D; S H C; S D C; H D C) and four ways in which a card of any suit can be selected. $4 \times 4 \times 4 \times 4$ is 256. Here the punter's chance is 256/560 or 16/35.

Hence a punter backing the appearance of three cards of a suit stands to win one in 35 attempts. Her "expectation" is a loss of 34 chips to gain 30. Net loss: 4 chips.

A punter backing the appearance of cards of three different suits stands to win 16 times out of 35. Net loss is 3 chips.

The latter is therefore the more advantageous bet.

60 BRACELETS

Experiment will quickly show that there are only three basic patterns. Call the beads R, G, and Y. Then we have:

(1)	R R G G Y	R R Y Y G	G G Y Y R		= 3
(2)	R G G R Y	R Y Y R G	G Y Y G R	⎫	
	G R R G Y	Y R R Y G	Y G G Y R	⎬ = 6	
(3)	R G R G Y	R Y R Y G	G Y G Y R		= 3

So there are 12 distinguishable bracelets in all.

61 SIX BOWLER HATS

The three hats can be taken from the hatstand in 120 different ways. For the first brother has 6 hats to choose from; the second has 5; the third has 4. And $6 \times 5 \times 4 = 120$.

How many of these 120 selections give each of the brothers a hat other than his own?

Call the brothers T D H. Call T's hats t, D's hats d, H's hats h. Then TDH can take respectively *dht* or *htd*. And since each brother has two hats, 8 selections give *dht* and 8 give *htd*.

But TDH may also get one of the following selections: *dhd; dtd; dtt; htt; hht; hhd*. There are only four of each of these: the two hats of one brother are taken together with either of the hats of a second.

Hence there are $16 + 24$, i.e., 40 selections of the three hats which give each brother a hat not his own.

So, as before, there is **one chance in three that none of the brothers takes one of his own hats.**

62 DARTS

This, if tackled the right way, is not at all a difficult problem.

Call the teams F C U B S.

On the first Saturday F play C. B must have played in the other match, for they have their bye in Round 4. But B play U in Round 2, so they play S in Round 1, and U have the bye. S cannot have the bye in Round 2; for, if so, F would meet C, whom they have played already. And S cannot have the bye in Round 3, for they are playing U.

So on the fifth Saturday **Sceptre have the bye.** This, incidentally, is the only fact which can be determined with certainty.

63 CHESTNUT AVENUE

Try adding all the numbers and dividing by six. The quotient is 273. Then a little thought suggests the following series:

(1)	1	16	256
(2)	3	27	243
(3)	13	52	208
(4)	21	63	189
(5)	28	70	175
(6)	39	78	156

In each series the total is 273, and the middle term is the geometric mean of the others.

64 THE LIARS

The lowest common multiple of 2, 3, and 5 is 30. We therefore need to analyze the 30 possibilities which arise when L, F, and S make one statement each. Putting T for a true statement, and L for a lie, this analysis works out as follows:

LIZARD:	T	T	T	L	T	L	L	L
FLITT:	T	T	L	T	L	T	L	L
STOAT:	T	L	T	T	L	L	T	L

2	4	3	2	6	4	3	6

But the statements TTT, TTL, TLT, LTT do not enter into our final calculations, since only one of the three can possibly be true. All three, of course, may be lies. Hence, there are 3 chances in 19 that Flittermouse is the winner (Stoat's statement), and 6 chances in 19 that all three statements are lies and that Puddock is the winner.

It is twice as likely that Puddock is the winner as that the winner is Flittermouse.

65 MR. STOUT AND MR. PORTER

Each coin is equally likely to come down "heads" or "tails." So there are $2 \times 2 \times 2 \times 2 \times 2$ (i.e., 32) possibilities when Stout throws five coins, and 32 possibilities when Porter throws them; and we have to consider 32×32 or 1,024 possibilities in all.

Analyzing the 32 possibilities arising from throwing five coins, it will be found that there is one chance of five heads and one chance of no heads; that there are five chances of four heads and five chances of one head; and that there are 10 chances of three heads and 10 chances of two heads.

So the chance that S and P will throw the same number of heads is $2(1 \times 1)$ plus $2(5 \times 5)$ plus $2(10 \times 10)$, i.e., 252/1,024.

Hence the chance that P pays for the drinks is $(252 + 386)/1,024$, or $638 : 386$.

66 SNOOKER

This is a very pretty problem. It fulfils the first condition of a good puzzle: that it isn't obvious how to set about it.

The simplest approach is this. Consider how many points in all are scored where a player wins a specified number of frames.

IF HE WINS	HE SCORES	AND OTHERS SCORE	TOTAL
4	4	0	4
3	3	$1\frac{1}{2}$	$4\frac{1}{2}$
2	2	2	4
1	1	$1\frac{1}{2}$	$2\frac{1}{2}$
0	0	0	0

Now 10 frames in all are played. Set out their possible distribution and the consequent aggregate of points:

(1)	4	3	2	1	0	15
(2)	4	2	2	2	0	16
(3)	4	2	2	1	1	17
(4)	3	3	3	1	0	16
(5)	3	3	2	2	0	17
(6)	3	3	2	1	1	18
(7)	3	2	2	2	1	19
(8)	2	2	2	2	2	20

So (6) is the relevant distribution, and it follows that **Green won three games.**

67 DIMONT WAS UNLUCKY

Dimont loses three rubbers; each of the other players wins two.

Call the stakes for the first rubber m dollars per 100; let the first rubber be n times 100 points. Then m and n are both integers, and n is greater than 6.

The first rubber costs the losers mn dollars each; the second rubber, $2m(n-3)$ dollars; the third rubber, $4m(n-6)$ dollars.

So, since Harz wins 25 dollars,

$$mn + 2m(n-3) - 4m(n-6) = 25$$
$$\text{i.e., } m(18-n) = 25$$

There are two integral solutions:

$$(1) \ m = 5; \qquad n = 13$$
$$(2) \ m = 25; \qquad n = 17$$

But (2) is inadmissible, since Dimont would lose $(425 + 700 + 1{,}100)$ dollars, i.e., over \$2,000.

Hence in fact Dimont loses \$65 plus \$100 plus \$140, or \$305 in all.

68 POSTAGE STAMPS

In all, 1,000 stamps are distributed. Suppose that, on Monday, $(m-4)$ applicants each receive $(n+20)$ stamps. Then we have:

$$(m-4)(n+20) + (m-2)(n+10) + mn + (m+2)$$
$$(n-10) + (m+4)(n-20) = 1,000$$

Whence $mn = 240$.

There are three possible values of m and n. These give, day by day, the following distributions:

$$(1) \quad m = 6; \qquad n = 40$$
$$(2) \quad m = 8; \qquad n = 30$$
$$(3) \quad m = 10; \qquad n = 24$$

	MON.	TUES.	WED.	THURS.	FRI.
(1)	120	200	240	240	200
(2)	200	240	240	200	120
(3)	264	272	240	168	56

So (3) is the relevant distribution. The number of applicants for stamps is $6+8+10+12+14$, i.e., **50 collectors took advantage of Squeers' offer.**

69 TWENTY-ONE RUBBERS

This is a simple exercise in numerical analysis. We have first to discover what number can be divided into three other numbers in exactly seven different ways. The rest will then follow.

The required number (arrived at by experiment) is 9. We can now construct the following table showing the results of each night's play. (We do not know in what order the several results occurred: but this is immaterial.)

SCORES OF 3 RUBBERS IN HUNDREDS			PLAYERS' WINS OR LOSSES			
			N	E	W	S
3	3	3	+9	−3	−3	−3
4	3	2	+9	−1	−3	−5
4	4	1	+9	−1	−1	−7
5	2	2	+9	+1	−5	−5
5	3	1	+9	+1	−3	−7
6	2	1	+9	+3	−5	−7
7	1	1	+9	+5	−7	−7
		Total	+63	+5	−27	−41

So North won \$63; East won \$5; West lost \$27; South lost \$41.

70 RUY LOPEZ

We may first analyze the possible distribution of fourteen players representing three schools:

DISTRIBUTION			NO. OF GAMES			TOTAL
A	B	C	A–B	A–C	B–C	
3	3	8	9	24	24	57
3	4	7	12	21	28	61
3	5	6	15	18	30	63
4	4	6	16	24	24	64
4	5	5	20	20	25	65

This shows at a glance that the relevant distribution is 3, 5, 6. B is Pawnborough, who play 9 Ruy Lopez games. C is Two Bishops, who play 12 Ruy Lopez games. As, from the point of view of the three schools, 42 Ruy Lopez games in all were played, Castleton must have taken part in them all.

So Castleton were concerned in no fewer than 21 Ruy Lopez games.

71 THERE AND BACK

Let Bob's outward speed be x miles an hour. He returns at $5x/4$ miles per hour.

Let the Colonel's outward speed be y miles an hour. He returns at $2y$ miles·an hour.

Then $40/x + 40/(5x/4) = 40/y + 40/2y$.

So $y = 5x/6$.

Hence, when Bob reaches Hightown, the Colonel is $40/6$ miles away. Now Bob increases his speed to $5x/4$ m.p.h. The two cars are approaching one another at $(5x/4 + 5x/6)$ m.p.h. So, when they meet, the Colonel has covered $2/5$ of the distance. He is 36 miles from home.

The Blue Griffin is four miles from the Market Cross at Hightown.

72 ISLE OF MUGS

1. Call the drivers G and S. We must first determine their relative speeds. If G does his first circuit at m miles per hour, we have $1/m + 1/(m+30) = 2/(m+12)$. So $m = 60$. G does his first circuit at 60 m.p.h. and his second circuit at 90 m.p.h.; and S does both circuits at 72 m.p.h.

2. Hence the relative distances covered before G and S meet for the first time are G, 5; S, 6. The relative distances covered after they have met for the third time are G, 5; S, 4. So, if t be the length of the circuit in miles, $5t/11 - 4t/9 = 1$. The circuit is 99 miles.

3. When G and S meet for the second time, therefore, S has covered 108 miles to G's 90 miles.

And the point where they meet is nine miles distant from the White Swan.

73 FUGITIVE FOUR

There are 3 possible results in respect of each match. So there are $3 \times 3 \times 3 \times 3$, i.e., 81 possible forecasts in all.

Make any assumption (it does not matter what) in regard to Black's forecast. E.g., suppose his dice give him H A H D.

Now consider White's 81 possible forecasts in relation to this forecast of Black's.

(*a*) White can get an identical forecast in one way only.

(*b*) He can get the same forecast in respect of three matches in 4×2, i.e., in 8 ways. For any one of the four matches may be the one in respect of which he differs; and for this match he can get two different results.

(*c*) Similarly, he can get the same forecast as Black in respect of two matches in $6 \times 2 \times 2$, i.e., in 24 ways;

(*d*) and he can get the same forecast as Black in respect of one match only in $4 \times 2 \times 2 \times 2$, i.e., in 32 ways;

(*e*) and, finally, he can get a completely different forecast from Black's in $2 \times 2 \times 2 \times 2$, i.e., in 16 ways.

To check these calculations: $1 + 8 + 24 + 32 + 16 = 81$. **So White's contention is correct.** There are 32 chances in 81 that the two forecasts will agree in respect of just one match, and only 16 chances in 81 that they will not agree at all.

74 FIREWORKS

Let A equal the number of rockets, and G's age.

$$\frac{A}{10} = \frac{A}{2} - \frac{A}{6} - 21$$

The right-hand side reduces to:

$$\frac{A}{3} - 21$$

Thus

$$21 = \frac{A}{3} - \frac{A}{10} = \frac{7A}{30}$$

So, $A = 90$, and Grandfather will be 100 in ten years' time.

75 MUNICIPAL PARK

We are clearly concerned with "solving" a right-angled triangle. Let its longer side be m yards; its shorter side, n yards; its diagonal, p yards.

Then (i) $n^2 = p^2 - m^2$
$$= (p-m)(p+m) \cdot$$
(ii) $n = 2(p+m)/7$
i.e., $n^2 = 4(p+m)(p+m)/49$
So $4(p+m) = 49(p-m)$
i.e., $45p = 53n$

and the sides of the right-angled triangle are in the ratio: 53 : 45 : 28. But the diagonal is 250 yards longer than the shorter side. Hence the sides of the rectangle are 280 yards and 450 yards, and **the area of the park is 126,000 square yards.**

76 GREENLEAF

Since "if you take a chance," the odds are 7 to 3 on the next train being a train for Greenleaf; and since there is one Greenleaf train, and one Blackthorn train, every half-hour, the Greenleaf train must leave 21 minutes after the train for Blackthorn.
So my train left for Greenleaf at 9:43.

77 WRECK OF THE *HESPERUS*

Assume that there are m women, and that each child receives $\$p$. Then, since there are twice as many men as women, we have the indeterminate equation:

$2m(6p+5) + 6mp + 20p = 1,000$
i.e., $p(18m+20) = 1,000 - 10m$

Since each child receives an exact number of dollars, we

must find a solution in integers. There are two such solutions:
(1) $m = 35$; $p = 1$. (2) $m = 9$; $p = 5$. The first of these is
excluded by the data.

So the money is distributed as follows:

20 children each receive	.	.	.	$5 = $100
9 women each receive	.	.	.	$30 = $270
18 men each receive	.	.	.	$35 = $630

and **there were 47 survivors in all.**

78 INSECT LEAGUE

There were six different scores, and only 11 goals in all. So
the six scores were: 3–0, 2–1, 2–0, 1–1, 1–0, 0–0.

Call the teams H, B, W, D.

Consider H. Their scores must have been 3–0, 0–0, 1–2, or
3–0, 1–1, 0–1. Now consider B. Their scores were either 2–1,
0–0, 0–1, or 1–0, 0–0, 1–2.

So the score in the game between H and B was (1) 0–0, or
(2) 1–2, or (3) 0–1.

Following up each of these in turn, we obtain a unique
solution:

AGAINST

	H	B	W	D
H	–	0	1	3
B	1	–	1	0
W	1	2	–	0
D	0	0	2	–

The Dragonflies beat the Wasps by 2 goals to 0.

79 BUTTER-FINGERS

This problem is not really very difficult.

Suppose that "B-F" had originally dropped x Clubs.

Then there are $(13 - x)$ other cards in his hand, and each of
them is equally likely to be any one of $(52 - x)$ cards, of which
the Clubs number $(13 - x)$.

So we have the equation:

$$\frac{x + \dfrac{(13-x)(13-x)}{(52-x)}}{13} = \frac{7}{16}$$

i.e., $39x = 156$

Whence $x = 4$.

So originally Butter-Fingers dropped four Clubs.

80 FAMILY SWEEPSTAKE

A strictly "mathematical" solution isn't possible, but the problem presents little difficulty if the relevant data are intelligently analyzed.

Suppose there are a Gambler boys; b Gambler girls; c Plunger boys; d Plunger girls.

Then (1) $b = d$.

$$(2)\ \frac{a(a+1)}{a+2b+c} = \tfrac{1}{5}$$

$$(3)\ \frac{c(c+1)}{a+2b+c} = \tfrac{1}{7}$$

This at once suggests that $a + 2b + c = 21$.

a is 7; b is 4; c is 6; d is 4.

So the odds required are $\dfrac{2(7 \times 4)}{21 \times 20}$

I.e., **13 to 2 against.**

81 GRANDPA WAS EIGHTY-NINE

This is clearly an exercise in factorization, on a somewhat experimental basis. The difference between m^2 and n^2 is, of course, $(m+n)(m-n)$; so we have to find a number which offers a sufficiency of factors—each pair being either both odd, or both even. 1,680 is found to fill the bill, and Grandpa's age (89) confirms the assumption that 1,680 is correct.

AGES

			AGES		
We find that 1,680 =	(1)	168×10	89	(G)	79 (g)
	(2)	140×12	76	(?B)	64 (?b)
	(3)	120×14	67	(P)	53 (j)
	(4)	84×20	52	(?Gu)	32 (cl)
	(5)	70×24	47	(?W)	23 (ma)
	(6)	60×28	44	(D)	16 (ci)
	(7)	56×30	43	(mu)	13 (J)
	(8)	42×40	41	(f)	1 (ba)

Grandpa is G; Grandma, g; Mr. Brown, B; Mrs. Brown, b; Pongo, P; Julia, j; Guy, Gu; Clara, cl; Will, W; Mary, ma; Dad, D; Cissie, ci; Mum, mu; James, J; Fritillary, f; and Baby, ba.

James is thirteen.

82 RATHER SILLY

Let the three ages be a, b, c.

Let h_1 be their present harmonic mean, and h_2 their harmonic mean in 5 years' time. Then we can assume that 13 and 11 are denominators in this equation:

$$\frac{n_2}{13} - \frac{n_1}{11} = \frac{810}{143}$$

where $$h_1 = \frac{n_1}{11} \text{ and } h_2 = \frac{n_2}{13}$$

By trial, the only satisfactory values of n_1 and n_2 are 90 and 180 respectively, and the three ages are: Baby, 5; Belinda, 10, **and Betty is 15.**

83 NEWS FROM CRAB'S BAY

M covers 78 miles $(52 - 15 + 52 - 11)$ while P covers 26 $(15 + 11)$. So M's speed is three times P's speed.

So when M left Prawnacre, P was 20 miles $(15+5)$ distant; i.e., he had covered 32 miles.

Hence, if P's speed is x m.p.h.:

$$\frac{32}{x} + \frac{104}{3x} = \frac{20}{3}$$

So P's speed is 10 m.p.h.

And Pedal was back at Crab's Bay at 8:24 P.M.

84 ESSAYS

Let there be A boys and B girls competing.

Let Mrs. McScribble's children be a boys and b girls.

Then $A+B = 20: A < B; a+b = 6$.

The chance of a boys winning all the prizes is

$$\frac{1}{^{A}C_{a}} \text{ if there are } A \text{ boys}$$

$$\text{or } \frac{1}{^{A+1}C_{a}} \text{ if there are } (A+1) \text{ boys.}$$

Similarly the chance of b girls winning is

$$\frac{1}{^{B}C_{b}} \text{ if there are } B \text{ girls}$$

$$\text{or } \frac{1}{^{B-1}C_{b}} \text{ if there are } (B-1) \text{ girls}$$

$$\text{Hence } \frac{^{A+1}C_{a}}{^{A}C_{a}} = \frac{^{B}C_{b}}{^{B-1}C_{b}}$$

By inspection, the only feasible solutions are:

   (1) $a = 2; b = 4$.   (2) $a = 4; b = 2$.

   (1) gives $A = 6, B = 14$;   (2) gives $A = 13, B = 7$.

But $A < B$, so (1) is a unique answer.

(1) There are 6 boys and 14 girls competing. (2) Mrs. McScribble has two sons and four daughters.

85 NARROW TRIUMPH FOR CLOWNCHESTER

We do not know how many events there were. We have to ascertain the point basis in each event.

Let m, n, p points respectively be scored for first, second, and third places.

Then m is greater than n; n is greater than p; and $4m + 2n + p = 25$. Clearly p is either 3 or 1.

(1) If p is 3, $2m + n = 11$. This is impossible.

(2) So p is 1. $2m + n = 12$. Clearly n is 2 and m is 5.

Hence (though we do not need to know this) there are 6 events.

And Peter Probe scored 12 points.

86 ALLOTMENTS

Clearly we have to find four different integral values of m and n which satisfy the expression: $m^2 + n^2 = 221^2$

Now $221 = 13 \times 17$

So $221^2 \quad = 221 \times 221$ (1)

$\qquad\qquad$ or 169×289 (2)

But $\begin{cases} (1) \ \ 221 = 11^2 + 10^2 \ (a) \text{ or } 14^2 + 5^2 \ (b) \\ (2) \ \ 169 = 12^2 + 5^2 \ (c) \\ (3) \ \ 289 = 15^2 + 8^2 \ (d) \end{cases}$

Hence from (a) we get $m = 220$, $n = 21$.

$\qquad$ „   (b)   „   „ $m = 171$, $n = 140$.

$\qquad$ „   (c)   „   „ $m = 204$, $n = 85$.

$\qquad$ „   (d)   „   „ $m = 195$, $n = 104$.

So the areas of the four allotments are: (a) 4,620 sq. ft.; (b) 23,940 sq. ft.; (c) 17,340 sq. ft.; (d) 20,280 sq. ft.

By inspection, (a) is D's; (b) is A's; (c) is B's; (d) is C's.

The area of Adam's allotment exceeds the area of Dave's by 19,320 sq. ft.

87 THE PROFESSOR'S DAUGHTER

The point of this puzzle is that the Colonel "never makes a mistake in arithmetic." So, since he has two bad shots, there must be at least three combinations of daughters' ages which, multiplied together, give 1,200, and, added together, give the same total.

It will be found (by trial) that the only ages conforming to the data are:

(1)	4	15	20
(2)	5	10	24
(3)	6	8	25

The total in each case is 39.

So the professor's wife is forty-nine.

88 JONES COLLECTED ONE DOLLAR

Jones had pulled a fast one. It is quite true (as can very quickly be ascertained) that 19 of 36 throws produce a product less than 12, and four a product of 12 exactly. But Jones omitted to mention that the mean of the 36 throws—the thrower's "expectation"—is in fact $12\frac{1}{4}$.

89 DEAD ON TIME

If Speedmore's cycling speed is m miles per hour, we have:

$$\frac{300}{m} + \frac{300}{3m} + \frac{300}{3m+15} = 24$$

i.e., $6m^2 - 95m - 500 = 0$

whence $(6m + 25)(m - 20) = 0$.

So Speedmore's cycling speed averaged 20 m.p.h.

90 CUTTING FOR PARTNERS

This puzzle answers a question which is often put to me.

The total number of combinations of four cards is $^{52}C_4$, i.e., 270,725.

This (perhaps surprisingly) large total can be analyzed as follows: There are five possible distributions amongst suits:

(1) 4 0 0 0   (2) 3 1 0 0   (3) 2 2 0 0   (4) 2 1 1 0
(5) 1 1 1 1

Their several frequencies can be calculated as under:

(1)	4 0 0 0	$4 \times {}^{13}C_4$	2,860
(2)	3 1 0 0	$4 \times {}^{13}C_3 \times 39$	44,616
(3)	2 2 0 0	$6 \times {}^{13}C_2 \times {}^{13}C_2$	36,504
(4)	2 1 1 0	$12 \times {}^{13}C_2 \times 13 \times 13$	158,184
(5)	1 1 1 1	$13 \times 13 \times 13 \times 13$	28,561
		TOTAL	270,725

So the odds in case (4) are 158,184 to 112,541; or roughly 10 to 7.

91 SAVINGS CERTIFICATES

Let x be the number of Sir Henry's nephews, and y the number of certificates which each nephew receives.

Then $xy + (x-3)(y-5) = 1,949$.

I.e., $2xy - 3y - 5x + 15 = 1,949$.

Whence $4xy - 6y - 10x + 15 = 3,898 - 15$.

Or $(2x-3)(2y-5) = 3,883 = 11 \times 353$.

But 11 and 353 are both prime numbers. So $x = 7$ and $y = 179$.

Each nephew receives 179 certificates.

92 MARGARINE

Suppose there are m pats of butter.

Then there are $m+2$ pats of margarine, and the total number of pats is $2m+2$.

If, then, Probe selects three pats at random, the chance that all three are margarine is

$$\frac{(m+2)\ C_3}{(2m+2)C_3}$$

And this equals $\frac{1}{6}$.

i.e., $\dfrac{(m+2)! \times (2m-1)!}{(2m+2)! \times (m-1)!} = \frac{1}{6}$

So $\dfrac{m(m+1)(m+2)}{2m(2m+1)(2m+2)} = \frac{1}{6}$

i.e., $m^2 - 3m - 4 = 0$

Whence $m = 4$.

There are 4 pats of butter and 6 pats of margarine, and the chance that Probe can select 3 butters at random is

$$\frac{{}^4C_3}{{}^{10}C_3}$$ i.e., **29 to 1 against.**

93 HEADACHE FOR THE CITY SURVEYOR

Let the present working day be m hours and the payment for a working day x dollars.

Then $x/m = 40/x$. i.e., $x^2 = 40m$.

Also the proposed working day is $(m-1)$ hours; and the proposed payment for a working day $(x+1)$ dollars.

So $(x+1)/(m-1) = 49/(x+1)$ i.e., $(x+1)^2 = 49(m-1)$.

Eliminating m, we get:

$9x^2 - 80x - 2{,}000 = 0$. i.e., $(x-20)(9x+100) = 0$

So x is 20; whence m is 10.

The present working day is thus 10 hours, and the hourly rate \$2 and **the proposed working day is 9 hours and proposed hourly rate \$2$\frac{1}{3}$.**

94 GRAND CHAIN

Alice's two dominoes totalling 23 must have been 6–6 and 6–5.
Betty's two dominoes totalling 20 must have been 6–4 and 5–5.
Clara's two dominoes totalling 18 must have been 6–3 and 5–4.
Dolly's two dominoes totalling 16 must have been 5–3 and 4–4,
since the 6–2 is in Alice's hand.

Alice cannot begin by playing the 6–6, for if she does, Betty
must play the 6–4, Clara, the 6–3, and Dolly either the 3–5
or the 4–4. No piece can be played after the 4–4, and only
one after the 3–5.

Therefore Alice must lead off with the 6–5, and the play is
necessarily as follows:

1st round:	**6-5**	**5-5**	**5-4**	**4-4**
2nd round:	**6-6**	**6-4**	**6-3**	**5-3**

95 TEDDY BEARS

Let either purchaser buy m bears on the second day at a cost
of n cents each.

Then his (or her) total expenditure (in cents) is:

$$(m+1)(n-20) + mn + (m-1)(n+20)$$
$$\text{i.e., } 3mn - 40 = 2{,}867$$
$$\text{whence } mn = 969.$$

Now the factors of 969 are 3, 17, 19.

And since " as to price, there wasn't a great deal in it," one
shopper bought 17 middling-sized bears at 57 cents each; the
other bought 19 at 51 cents each. The latter is clearly Betty.

Purchases on the three days are therefore:

		BONIFACE			BETTY		
				Cents			*Cents*
1st day	. .	18 bears at 37¢		666	18 at 71¢		1,278
2nd ,,	. .	17 ,, ,, 57¢		969	19 ,, 51¢		969
3rd ,,	. .	16 ,, ,, 77¢		1,232	20 ,, 31¢		620
TOTALS	. .	51 bears		2,867	57 bears		2,867

So Betty bought six bears more than Boniface.

96 RIPCHESTER

This is one of my "cryptic" problems: it's hard to see how there can be a unique solution.

Consider, however, the nine "placings." The digits 1 2 3 . . . 9 must be divided into three groups of three each which add up to the same total. Reflection will show that there are only two possibilities:

(1)	1	5	9	2	6	7	3	4	8
(2)	1	6	8	2	4	9	3	5	7

In either case, the three groups represent the placings of the three houses. Clearly (1) alone is relevant; George and James were respectively 6th and 7th, and

Tom finished second.

97 A LOSING BATTLE

Suppose that, on the fifth day, each of m men shifts n crates. The total is mn. The day before the total was $(m-6)(n+5)$; the day after $(m+6)(n-5)$; and so on. Adding these totals, we get:

$$9mn - 1,800 = 69,489$$
$$\text{whence } mn = 7,921.$$

This number (7,921) is the square of 89, a prime number; so both m and n are 89; and we can now construct the table of work done each day:

1st day:	65 men each shift 109 crates: 7,085
2nd ,, :	71 ,, ,, ,, 104 ,, : 7,384
3rd ,, :	77 ,, ,, ,, 99 ,, : 7,623
4th ,, :	83 ,, ,, ,, 94 ,, : 7,802
5th ,, :	89 ,, ,, ,, 89 ,, : 7,921
6th ,, :	95 ,, ,, ,, 84 ,, : 7,980
7th ,, :	101 ,, ,, ,, 79 ,, : 7,979
8th ,, :	107 ,, ,, ,, 74 ,, : 7,918
9th ,, :	113 ,, ,, ,, 69 ,, : 7,797

So the largest number of crates shifted in one day was 7,980 (sixth day).

98 THE THIRTY-SEVENTH INTERNATIONAL

The clue is "Thirty-seventh." If 1/37, 2/37, and so on up to 36/37 are written as recurring decimals, we get twelve groups of 3 digits, each group arranged in cyclic order.

They are:

1	2	3	4	5	6	7	8	9	10	11	12
027	054	081	108	135	162	189	216	243	270	297	324

13	14	15	16	17	18	19	20	21	22	23	24
351	378	405	432	459	486	513	540	567	594	621	648

25	26	27	28	29	30	31	32	33	34	35	36
675	702	729	756	783	810	837	864	891	918	945	972

So Number Three's colleagues are No. 4 and No. 30.

99 THE INGENIOUS MR. DRAKE

This is not difficult.

Let there be x competitors in Section A.

Then there are $(x+2)$ competitors in Section B,

and $(28-2x)$ competitors in Section C.

$$\text{So } \frac{x(x-1)}{2} + \frac{(x+2)(x+1)}{2} = \frac{(28-2x)(27-2x)}{2} + 7$$

This resolves itself into the quadratic:

$$x^2 - 56x + 384 = 0.$$

Whence $(x-8)(x-48) = 0.$

I.e., $x = 8.$

So there are 8, 10, and 12 competitors respectively in Sections A, B, C.

100 THE LOCAL BOTWINNIKS

Suppose that last year there were m sections, with n players in each.

Then $(mn - 2) = (m + 1)(n - 1)$

i.e., $m = (n + 1)$

So $(n + 1) \times n(n - 1) = (n + 2)(n - 1)(n - 2)$

This leads to the equation

$(n - 10)(n + 13) = 0,$

whence $n = 10$.

So last year there were 11 sections of 10 players each, and this year **there are 12 sections of 9 players each, or 108 competitors in all.**

A CATALOGUE OF SELECTED DOVER BOOKS
IN ALL FIELDS OF INTEREST

A CATALOGUE OF SELECTED DOVER BOOKS
IN ALL FIELDS OF INTEREST

AMERICA'S OLD MASTERS, James T. Flexner. Four men emerged unexpectedly from provincial 18th century America to leadership in European art: Benjamin West, J. S. Copley, C. R. Peale, Gilbert Stuart. Brilliant coverage of lives and contributions. Revised, 1967 edition. 69 plates. 365pp. of text.

21806-6 Paperbound $3.00

FIRST FLOWERS OF OUR WILDERNESS: AMERICAN PAINTING, THE COLONIAL PERIOD, James T. Flexner. Painters, and regional painting traditions from earliest Colonial times up to the emergence of Copley, West and Peale Sr., Foster, Gustavus Hesselius, Feke, John Smibert and many anonymous painters in the primitive manner. Engaging presentation, with 162 illustrations. xxii + 368pp.

22180-6 Paperbound $3.50

THE LIGHT OF DISTANT SKIES: AMERICAN PAINTING, 1760-1835, James T. Flexner. The great generation of early American painters goes to Europe to learn and to teach: West, Copley, Gilbert Stuart and others. Allston, Trumbull, Morse; also contemporary American painters—primitives, derivatives, academics—who remained in America. 102 illustrations. xiii + 306pp. 22179-2 Paperbound $3.00

A HISTORY OF THE RISE AND PROGRESS OF THE ARTS OF DESIGN IN THE UNITED STATES, William Dunlap. Much the richest mine of information on early American painters, sculptors, architects, engravers, miniaturists, etc. The only source of information for scores of artists, the major primary source for many others. Unabridged reprint of rare original 1834 edition, with new introduction by James T. Flexner, and 394 new illustrations. Edited by Rita Weiss. 6⅝ x 9⅝.

21695-0, 21696-9, 21697-7 Three volumes, Paperbound $13.50

EPOCHS OF CHINESE AND JAPANESE ART, Ernest F. Fenollosa. From primitive Chinese art to the 20th century, thorough history, explanation of every important art period and form, including Japanese woodcuts; main stress on China and Japan, but Tibet, Korea also included. Still unexcelled for its detailed, rich coverage of cultural background, aesthetic elements, diffusion studies, particularly of the historical period. 2nd, 1913 edition. 242 illustrations. lii + 439pp. of text.

20364-6, 20365-4 Two volumes, Paperbound $6.00

THE GENTLE ART OF MAKING ENEMIES, James A. M. Whistler. Greatest wit of his day deflates Oscar Wilde, Ruskin, Swinburne; strikes back at inane critics, exhibitions, art journalism; aesthetics of impressionist revolution in most striking form. Highly readable classic by great painter. Reproduction of edition designed by Whistler. Introduction by Alfred Werner. xxxvi + 334pp.

21875-9 Paperbound $2.50

DESIGN BY ACCIDENT; A BOOK OF "ACCIDENTAL EFFECTS" FOR ARTISTS AND DESIGNERS, James F. O'Brien. Create your own unique, striking, imaginative effects by "controlled accident" interaction of materials: paints and lacquers, oil and water based paints, splatter, crackling materials, shatter, similar items. Everything you do will be different; first book on this limitless art, so useful to both fine artist and commercial artist. Full instructions. 192 plates showing "accidents," 8 in color. viii + 215pp. 8⅜ x 11¼.   21942-9 Paperbound $3.50

THE BOOK OF SIGNS, Rudolf Koch. Famed German type designer draws 493 beautiful symbols: religious, mystical, alchemical, imperial, property marks, runes, etc. Remarkable fusion of traditional and modern. Good for suggestions of timelessness, smartness, modernity. Text. vi + 104pp. 6⅛ x 9¼.
                    20162-7 Paperbound $1.25

HISTORY OF INDIAN AND INDONESIAN ART, Ananda K. Coomaraswamy. An unabridged republication of one of the finest books by a great scholar in Eastern art. Rich in descriptive material, history, social backgrounds; Sunga reliefs, Rajput paintings, Gupta temples, Burmese frescoes, textiles, jewelry, sculpture, etc. 400 photos. viii + 423pp. 6⅜ x 9¾.   21436-2 Paperbound $4.00

PRIMITIVE ART, Franz Boas. America's foremost anthropologist surveys textiles, ceramics, woodcarving, basketry, metalwork, etc.; patterns, technology, creation of symbols, style origins. All areas of world, but very full on Northwest Coast Indians. More than 350 illustrations of baskets, boxes, totem poles, weapons, etc. 378 pp.
                    20025-6 Paperbound $3.00

THE GENTLEMAN AND CABINET MAKER'S DIRECTOR, Thomas Chippendale. Full reprint (third edition, 1762) of most influential furniture book of all time, by master cabinetmaker. 200 plates, illustrating chairs, sofas, mirrors, tables, cabinets, plus 24 photographs of surviving pieces. Biographical introduction by N. Bienenstock. vi + 249pp. 9⅞ x 12¾.   21601-2 Paperbound $4.00

AMERICAN ANTIQUE FURNITURE, Edgar G. Miller, Jr. The basic coverage of all American furniture before 1840. Individual chapters cover type of furniture—clocks, tables, sideboards, etc.—chronologically, with inexhaustible wealth of data. More than 2100 photographs, all identified, commented on. Essential to all early American collectors. Introduction by H. E. Keyes. vi + 1106pp. 7⅞ x 10¾.
                    21599-7, 21600-4 Two volumes, Paperbound $11.00

PENNSYLVANIA DUTCH AMERICAN FOLK ART, Henry J. Kauffman. 279 photos, 28 drawings of tulipware, Fraktur script, painted tinware, toys, flowered furniture, quilts, samplers, hex signs, house interiors, etc. Full descriptive text. Excellent for tourist, rewarding for designer, collector. Map. 146pp. 7⅞ x 10¾.
                    21205-X Paperbound $2.50

EARLY NEW ENGLAND GRAVESTONE RUBBINGS, Edmund V. Gillon, Jr. 43 photographs, 226 carefully reproduced rubbings show heavily symbolic, sometimes macabre early gravestones, up to early 19th century. Remarkable early American primitive art, occasionally strikingly beautiful; always powerful. Text. xxvi + 207pp. 8⅜ x 11¼.   21380-3 Paperbound $3.50

JOHANN SEBASTIAN BACH, Philipp Spitta. One of the great classics of musicology, this definitive analysis of Bach's music (and life) has never been surpassed. Lucid, nontechnical analyses of hundreds of pieces (30 pages devoted to St. Matthew Passion, 26 to B Minor Mass). Also includes major analysis of 18th-century music. 450 musical examples. 40-page musical supplement. Total of xx + 1799pp.
(EUK) 22278-0, 22279-9 Two volumes, Clothbound $15.00

MOZART AND HIS PIANO CONCERTOS, Cuthbert Girdlestone. The only full-length study of an important area of Mozart's creativity. Provides detailed analyses of all 23 concertos, traces inspirational sources. 417 musical examples. Second edition. 509pp.
(USO) 21271-8 Paperbound $3.50

THE PERFECT WAGNERITE: A COMMENTARY ON THE NIBLUNG'S RING, George Bernard Shaw. Brilliant and still relevant criticism in remarkable essays on Wagner's Ring cycle, Shaw's ideas on political and social ideology behind the plots, role of Leitmotifs, vocal requisites, etc. Prefaces. xxi + 136pp.
21707-8 Paperbound $1.50

DON GIOVANNI, W. A. Mozart. Complete libretto, modern English translation; biographies of composer and librettist; accounts of early performances and critical reaction. Lavishly illustrated. All the material you need to understand and appreciate this great work. Dover Opera Guide and Libretto Series; translated and introduced by Ellen Bleiler. 92 illustrations. 209pp.
21134-7 Paperbound $1.50

HIGH FIDELITY SYSTEMS: A LAYMAN'S GUIDE, Roy F. Allison. All the basic information you need for setting up your own audio system: high fidelity and stereo record players, tape records, F.M. Connections, adjusting tone arm, cartridge, checking needle alignment, positioning speakers, phasing speakers, adjusting hums, trouble-shooting, maintenance, and similar topics. Enlarged 1965 edition. More than 50 charts, diagrams, photos. iv + 91pp.
21514-8 Paperbound $1.25

REPRODUCTION OF SOUND, Edgar Villchur. Thorough coverage for laymen of high fidelity systems, reproducing systems in general, needles, amplifiers, preamps, loudspeakers, feedback, explaining physical background. "A rare talent for making technicalities vividly comprehensible," R. Darrell, *High Fidelity*. 69 figures. iv + 92pp.
21515-6 Paperbound $1.00

HEAR ME TALKIN' TO YA: THE STORY OF JAZZ AS TOLD BY THE MEN WHO MADE IT, Nat Shapiro and Nat Hentoff. Louis Armstrong, Fats Waller, Jo Jones, Clarence Williams, Billy Holiday, Duke Ellington, Jelly Roll Morton and dozens of other jazz greats tell how it was in Chicago's South Side, New Orleans, depression Harlem and the modern West Coast as jazz was born and grew. xvi + 429pp.
21726-4 Paperbound $2.50

FABLES OF AESOP, translated by Sir Roger L'Estrange. A reproduction of the very rare 1931 Paris edition; a selection of the most interesting fables, together with 50 imaginative drawings by Alexander Calder. v + 128pp. 6½x9¼.
21780-9 Paperbound $1.25

POEMS OF ANNE BRADSTREET, edited with an introduction by Robert Hutchinson. A new selection of poems by America's first poet and perhaps the first significant woman poet in the English language. 48 poems display her development in works of considerable variety—love poems, domestic poems, religious meditations, formal elegies, "quaternions," etc. Notes, bibliography. viii + 222pp.
22160-1 Paperbound $2.00

THREE GOTHIC NOVELS: THE CASTLE OF OTRANTO BY HORACE WALPOLE; VATHEK BY WILLIAM BECKFORD; THE VAMPYRE BY JOHN POLIDORI, WITH FRAGMENT OF A NOVEL BY LORD BYRON, edited by E. F. Bleiler. The first Gothic novel, by Walpole; the finest Oriental tale in English, by Beckford; powerful Romantic supernatural story in versions by Polidori and Byron. All extremely important in history of literature; all still exciting, packed with supernatural thrills, ghosts, haunted castles, magic, etc. xl + 291pp.
21232-7 Paperbound $2.00

THE BEST TALES OF HOFFMANN, E. T. A. Hoffmann. 10 of Hoffmann's most important stories, in modern re-editings of standard translations: Nutcracker and the King of Mice, Signor Formica, Automata, The Sandman, Rath Krespel, The Golden Flowerpot, Master Martin the Cooper, The Mines of Falun, The King's Betrothed, A New Year's Eve Adventure. 7 illustrations by Hoffmann. Edited by E. F. Bleiler. xxxix + 419pp.
21793-0 Paperbound $2.50

GHOST AND HORROR STORIES OF AMBROSE BIERCE, Ambrose Bierce. 23 strikingly modern stories of the horrors latent in the human mind: The Eyes of the Panther, The Damned Thing, An Occurrence at Owl Creek Bridge, An Inhabitant of Carcosa, etc., plus the dream-essay, Visions of the Night. Edited by E. F. Bleiler. xxii + 199pp.
20767-6 Paperbound $1.50

BEST GHOST STORIES OF J. S. LEFANU, J. Sheridan LeFanu. Finest stories by Victorian master often considered greatest supernatural writer of all. Carmilla, Green Tea, The Haunted Baronet, The Familiar, and 12 others. Most never before available in the U. S. A. Edited by E. F. Bleiler. 8 illustrations from Victorian publications. xvii + 467pp.
20415-4 Paperbound $3.00

THE TIME STREAM, THE GREATEST ADVENTURE, AND THE PURPLE SAPPHIRE— THREE SCIENCE FICTION NOVELS, John Taine (Eric Temple Bell). Great American mathematician was also foremost science fiction novelist of the 1920's. *The Time Stream*, one of all-time classics, uses concepts of circular time; *The Greatest Adventure*, incredibly ancient biological experiments from Antarctica threaten to escape; The *Purple Sapphire*, superscience, lost races in Central Tibet, survivors of the Great Race. 4 illustrations by Frank R. Paul. v + 532pp.
21180-0 Paperbound $3.00

SEVEN SCIENCE FICTION NOVELS, H. G. Wells. The standard collection of the great novels. Complete, unabridged. *First Men in the Moon, Island of Dr. Moreau, War of the Worlds, Food of the Gods, Invisible Man, Time Machine, In the Days of the Comet*. Not only science fiction fans, but every educated person owes it to himself to read these novels. 1015pp.
20264-X Clothbound $5.00

A History of Costume, Carl Köhler. Definitive history, based on surviving pieces of clothing primarily, and paintings, statues, etc. secondarily. Highly readable text, supplemented by 594 illustrations of costumes of the ancient Mediterranean peoples, Greece and Rome, the Teutonic prehistoric period; costumes of the Middle Ages, Renaissance, Baroque, 18th and 19th centuries. Clear, measured patterns are provided for many clothing articles. Approach is practical throughout. Enlarged by Emma von Sichart. 464pp. 21030-8 Paperbound $3.50

Oriental Rugs, Antique and Modern, Walter A. Hawley. A complete and authoritative treatise on the Oriental rug—where they are made, by whom and how, designs and symbols, characteristics in detail of the six major groups, how to distinguish them and how to buy them. Detailed technical data is provided on periods, weaves, warps, wefts, textures, sides, ends and knots, although no technical background is required for an understanding. 11 color plates, 80 halftones, 4 maps. vi + 320pp. 6⅛ x 9⅛. 22366-3 Paperbound $5.00

Ten Books on Architecture, Vitruvius. By any standards the most important book on architecture ever written. Early Roman discussion of aesthetics of building, construction methods, orders, sites, and every other aspect of architecture has inspired, instructed architecture for about 2,000 years. Stands behind Palladio, Michelangelo, Bramante, Wren, countless others. Definitive Morris H. Morgan translation. 68 illustrations. xii + 331pp. 20645-9 Paperbound $2.50

The Four Books of Architecture, Andrea Palladio. Translated into every major Western European language in the two centuries following its publication in 1570, this has been one of the most influential books in the history of architecture. Complete reprint of the 1738 Isaac Ware edition. New introduction by Adolf Placzek, Columbia Univ. 216 plates. xxii + 110pp. of text. 9½ x 12¾. 21308-0 Clothbound $10.00

Sticks and Stones: A Study of American Architecture and Civilization, Lewis Mumford.One of the great classics of American cultural history. American architecture from the medieval-inspired earliest forms to the early 20th century; evolution of structure and style, and reciprocal influences on environment. 21 photographic illustrations. 238pp. 20202-X Paperbound $2.00

The American Builder's Companion, Asher Benjamin. The most widely used early 19th century architectural style and source book, for colonial up into Greek Revival periods. Extensive development of geometry of carpentering, construction of sashes, frames, doors, stairs; plans and elevations of domestic and other buildings. Hundreds of thousands of houses were built according to this book, now invaluable to historians, architects, restorers, etc. 1827 edition. 59 plates. 114pp. 7⅞ x 10¾. 22236-5 Paperbound $3.00

Dutch Houses in the Hudson Valley Before 1776, Helen Wilkinson Reynolds. The standard survey of the Dutch colonial house and outbuildings, with constructional features, decoration, and local history associated with individual homesteads. Introduction by Franklin D. Roosevelt. Map. 150 illustrations. 469pp. 6⅝ x 9¼. 21469-9 Paperbound $4.00

ALPHABETS AND ORNAMENTS, Ernst Lehner. Well-known pictorial source for decorative alphabets, script examples, cartouches, frames, decorative title pages, calligraphic initials, borders, similar material. 14th to 19th century, mostly European. Useful in almost any graphic arts designing, varied styles. 750 illustrations. 256pp. 7 x 10.   21905-4 Paperbound $4.00

PAINTING: A CREATIVE APPROACH, Norman Colquhoun. For the beginner simple guide provides an instructive approach to painting: major stumbling blocks for beginner; overcoming them, technical points; paints and pigments; oil painting; watercolor and other media and color. New section on "plastic" paints. Glossary. Formerly *Paint Your Own Pictures*. 221pp.   22000-1 Paperbound $1.75

THE ENJOYMENT AND USE OF COLOR, Walter Sargent. Explanation of the relations between colors themselves and between colors in nature and art, including hundreds of little-known facts about color values, intensities, effects of high and low illumination, complementary colors. Many practical hints for painters, references to great masters. 7 color plates, 29 illustrations. x + 274pp.
20944-X Paperbound $2.50

THE NOTEBOOKS OF LEONARDO DA VINCI, compiled and edited by Jean Paul Richter. 1566 extracts from original manuscripts reveal the full range of Leonardo's versatile genius: all his writings on painting, sculpture, architecture, anatomy, astronomy, geography, topography, physiology, mining, music, etc., in both Italian and English, with 186 plates of manuscript pages and more than 500 additional drawings. Includes studies for the Last Supper, the lost Sforza monument, and other works. Total of xlvii + 866pp. 7⅞ x 10¾.
22572-0, 22573-9 Two volumes, Paperbound $10.00

MONTGOMERY WARD CATALOGUE OF 1895. Tea gowns, yards of flannel and pillow-case lace, stereoscopes, books of gospel hymns, the New Improved Singer Sewing Machine, side saddles, milk skimmers, straight-edged razors, high-button shoes, spittoons, and on and on . . . listing some 25,000 items, practically all illustrated. Essential to the shoppers of the 1890's, it is our truest record of the spirit of the period. Unaltered reprint of Issue No. 57, Spring and Summer 1895. Introduction by Boris Emmet. Innumerable illustrations. xiii + 624pp. 8½ x 11⅝.
22377-9 Paperbound $6.95

THE CRYSTAL PALACE EXHIBITION ILLUSTRATED CATALOGUE (LONDON, 1851). One of the wonders of the modern world—the Crystal Palace Exhibition in which all the nations of the civilized world exhibited their achievements in the arts and sciences—presented in an equally important illustrated catalogue. More than 1700 items pictured with accompanying text—ceramics, textiles, cast-iron work, carpets, pianos, sleds, razors, wall-papers, billiard tables, beehives, silverware and hundreds of other artifacts—represent the focal point of Victorian culture in the Western World. Probably the largest collection of Victorian decorative art ever assembled— indispensable for antiquarians and designers. Unabridged republication of the Art-Journal Catalogue of the Great Exhibition of 1851, with all terminal essays. New introduction by John Gloag, F.S.A. xxxiv + 426pp. 9 x 12.
22503-8 Paperbound $4.50

VISUAL ILLUSIONS: THEIR CAUSES, CHARACTERISTICS, AND APPLICATIONS, Matthew Luckiesh. Thorough description and discussion of optical illusion, geometric and perspective, particularly; size and shape distortions, illusions of color, of motion; natural illusions; use of illusion in art and magic, industry, etc. Most useful today with op art, also for classical art. Scores of effects illustrated. Introduction by William H. Ittleson. 100 illustrations. xxi + 252pp.

21530-X Paperbound $2.00

A HANDBOOK OF ANATOMY FOR ART STUDENTS, Arthur Thomson. Thorough, virtually exhaustive coverage of skeletal structure, musculature, etc. Full text, supplemented by anatomical diagrams and drawings and by photographs of undraped figures. Unique in its comparison of male and female forms, pointing out differences of contour, texture, form. 211 figures, 40 drawings, 86 photographs. xx + 459pp. 5⅜ x 8⅜.

21163-0 Paperbound $3.50

150 MASTERPIECES OF DRAWING, Selected by Anthony Toney. Full page reproductions of drawings from the early 16th to the end of the 18th century, all beautifully reproduced: Rembrandt, Michelangelo, Dürer, Fragonard, Urs, Graf, Wouwerman, many others. First-rate browsing book, model book for artists. xviii + 150pp. 8⅜ x 11¼.

21032-4 Paperbound $2.50

THE LATER WORK OF AUBREY BEARDSLEY, Aubrey Beardsley. Exotic, erotic, ironic masterpieces in full maturity: Comedy Ballet, Venus and Tannhauser, Pierrot, Lysistrata, Rape of the Lock, Savoy material, Ali Baba, Volpone, etc. This material revolutionized the art world, and is still powerful, fresh, brilliant. With *The Early Work,* all Beardsley's finest work. 174 plates, 2 in color. xiv + 176pp. 8⅛ x 11.

21817-1 Paperbound $3.00

DRAWINGS OF REMBRANDT, Rembrandt van Rijn. Complete reproduction of fabulously rare edition by Lippmann and Hofstede de Groot, completely reedited, updated, improved by Prof. Seymour Slive, Fogg Museum. Portraits, Biblical sketches, landscapes, Oriental types, nudes, episodes from classical mythology—All Rembrandt's fertile genius. Also selection of drawings by his pupils and followers. "Stunning volumes," *Saturday Review.* 550 illustrations. lxxviii + 552pp. 9⅛ x 12¼.

21485-0, 21486-9 Two volumes, Paperbound $7.00

THE DISASTERS OF WAR, Francisco Goya. One of the masterpieces of Western civilization—83 etchings that record Goya's shattering, bitter reaction to the Napoleonic war that swept through Spain after the insurrection of 1808 and to war in general. Reprint of the first edition, with three additional plates from Boston's Museum of Fine Arts. All plates facsimile size. Introduction by Philip Hofer, Fogg Museum. v + 97pp. 9⅜ x 8¼.

21872-4 Paperbound $2.00

GRAPHIC WORKS OF ODILON REDON. Largest collection of Redon's graphic works ever assembled: 172 lithographs, 28 etchings and engravings, 9 drawings. These include some of his most famous works. All the plates from *Odilon Redon: oeuvre graphique complet,* plus additional plates. New introduction and caption translations by Alfred Werner. 209 illustrations. xxvii + 209pp. 9⅛ x 12¼.

21966-8 Paperbound $4.00

MATHEMATICAL PUZZLES FOR BEGINNERS AND ENTHUSIASTS, Geoffrey Mott-Smith. 189 puzzles from easy to difficult—involving arithmetic, logic, algebra, properties of digits, probability, etc.—for enjoyment and mental stimulus. Explanation of mathematical principles behind the puzzles. 135 illustrations. viii + 248pp.

20198-8 Paperbound $1.25

PAPER FOLDING FOR BEGINNERS, William D. Murray and Francis J. Rigney. Easiest book on the market, clearest instructions on making interesting, beautiful origami. Sail boats, cups, roosters, frogs that move legs, bonbon boxes, standing birds, etc. 40 projects; more than 275 diagrams and photographs. 94pp.

20713-7 Paperbound $1.00

TRICKS AND GAMES ON THE POOL TABLE, Fred Herrmann. 79 tricks and games—some solitaires, some for two or more players, some competitive games—to entertain you between formal games. Mystifying shots and throws, unusual caroms, tricks involving such props as cork, coins, a hat, etc. Formerly *Fun on the Pool Table*. 77 figures. 95pp.

21814-7 Paperbound $1.00

HAND SHADOWS TO BE THROWN UPON THE WALL: A SERIES OF NOVEL AND AMUSING FIGURES FORMED BY THE HAND, Henry Bursill. Delightful picturebook from great-grandfather's day shows how to make 18 different hand shadows: a bird that flies, duck that quacks, dog that wags his tail, camel, goose, deer, boy, turtle, etc. Only book of its sort. vi + 33pp. 6½ x 9¼. 21779-5 Paperbound $1.00

WHITTLING AND WOODCARVING, E. J. Tangerman. 18th printing of best book on market. "If you can cut a potato you can carve" toys and puzzles, chains, chessmen, caricatures, masks, frames, woodcut blocks, surface patterns, much more. Information on tools, woods, techniques. Also goes into serious wood sculpture from Middle Ages to present, East and West. 464 photos, figures. x + 293pp.

20965-2 Paperbound $2.00

HISTORY OF PHILOSOPHY, Julián Marias. Possibly the clearest, most easily followed, best planned, most useful one-volume history of philosophy on the market; neither skimpy nor overfull. Full details on system of every major philosopher and dozens of less important thinkers from pre-Socratics up to Existentialism and later. Strong on many European figures usually omitted. Has gone through dozens of editions in Europe. 1966 edition, translated by Stanley Appelbaum and Clarence Strowbridge. xviii + 505pp.

21739-6 Paperbound $3.00

YOGA: A SCIENTIFIC EVALUATION, Kovoor T. Behanan. Scientific but non-technical study of physiological results of yoga exercises; done under auspices of Yale U. Relations to Indian thought, to psychoanalysis, etc. 16 photos. xxiii + 270pp.

20505-3 Paperbound $2.50

Prices subject to change without notice.
Available at your book dealer or write for free catalogue to Dept. GI, Dover Publications, Inc., 180 Varick St., N. Y., N. Y. 10014. Dover publishes more than 150 books each year on science, elementary and advanced mathematics, biology, music, art, literary history, social sciences and other areas.

SUMMER 77

INVENTORY 1983